Michele Viti

Precise and Fast Beam Energy Measurement

Michele Viti

Precise and Fast Beam Energy Measurement

Studies on the upstream beam energy monitors at the International Linear Collider

Südwestdeutscher Verlag für Hochschulschriften

Impressum/Imprint (nur für Deutschland/ only for Germany)
Bibliografische Information der Deutschen Nationalbibliothek: Die Deutsche Nationalbibliothek verzeichnet diese Publikation in der Deutschen Nationalbibliografie; detaillierte bibliografische Daten sind im Internet über http://dnb.d-nb.de abrufbar.

Alle in diesem Buch genannten Marken und Produktnamen unterliegen warenzeichen-, marken- oder patentrechtlichem Schutz bzw. sind Warenzeichen oder eingetragene Warenzeichen der jeweiligen Inhaber. Die Wiedergabe von Marken, Produktnamen, Gebrauchsnamen, Handelsnamen, Warenbezeichnungen u.s.w. in diesem Werk berechtigt auch ohne besondere Kennzeichnung nicht zu der Annahme, dass solche Namen im Sinne der Warenzeichen- und Markenschutzgesetzgebung als frei zu betrachten wären und daher von jedermann benutzt werden dürften.

Verlag: Südwestdeutscher Verlag für Hochschulschriften Aktiengesellschaft & Co. KG
Dudweiler Landstr. 99, 66123 Saarbrücken, Deutschland
Telefon +49 681 37 20 271-1, Telefax +49 681 37 20 271-0
Email: info@svh-verlag.de
Zugl.: Berlin, HU, Diss., 2010

Herstellung in Deutschland:
Schaltungsdienst Lange o.H.G., Berlin
Books on Demand GmbH, Norderstedt
Reha GmbH, Saarbrücken
Amazon Distribution GmbH, Leipzig
ISBN: 978-3-8381-1741-6

Imprint (only for USA, GB)
Bibliographic information published by the Deutsche Nationalbibliothek: The Deutsche Nationalbibliothek lists this publication in the Deutsche Nationalbibliografie; detailed bibliographic data are available in the Internet at http://dnb.d-nb.de.

Any brand names and product names mentioned in this book are subject to trademark, brand or patent protection and are trademarks or registered trademarks of their respective holders. The use of brand names, product names, common names, trade names, product descriptions etc. even without a particular marking in this works is in no way to be construed to mean that such names may be regarded as unrestricted in respect of trademark and brand protection legislation and could thus be used by anyone.

Publisher: Südwestdeutscher Verlag für Hochschulschriften Aktiengesellschaft & Co. KG
Dudweiler Landstr. 99, 66123 Saarbrücken, Germany
Phone +49 681 37 20 271-1, Fax +49 681 37 20 271-0
Email: info@svh-verlag.de

Printed in the U.S.A.
Printed in the U.K. by (see last page)
ISBN: 978-3-8381-1741-6

Copyright © 2010 by the author and Südwestdeutscher Verlag für Hochschulschriften Aktiengesellschaft & Co. KG and licensors
All rights reserved. Saarbrücken 2010

Contents

List of Figures   vii

List of Tables   ix

Introduction   xi

1 The International Linear Collider   **1**
   1.1 ILC Basic Design 1
      1.1.1 Electron Source 1
      1.1.2 Positron Source 2
      1.1.3 Damping Rings 3
      1.1.4 Main Linacs 4
      1.1.5 Beam Delivery System 5
   1.2 Physics at the ILC 7

2 Beam Energy Measurement Techniques   **11**
   2.1 Review on Methods 11
      2.1.1 Resonant Depolarization 11
      2.1.2 Compton Backscattering 13
      2.1.3 Deflection in a Dipole Field 14
      2.1.4 Radiative Return Events 17
   2.2 Energy Measurements in the Past 18
      2.2.1 BESSY I and II 18
      2.2.2 VEPP-4M 20
      2.2.3 Stanford Linear Collider 21
      2.2.4 Large Electron Positron Collider 22
   2.3 Beam Energy Measurement at the ILC 23

3 Magnetic Chicane as Beam Energy Spectrometer   **25**
   3.1 General Considerations 25
   3.2 SLAC Linac and End Station A 27
      3.2.1 SLAC Linac 27
      3.2.2 End Station A 27
   3.3 Experiment T474/491 29

	3.4	Resonant Cavity Beam Position Monitor	31
		3.4.1 Resonant Cavity and Beam Coupling	31
		3.4.2 Signal Processing	36

4 Characterization of the Magnets   41

    4.1   General Considerations 41
    4.2   The Magnets 10D37 41
    4.3   B-field Measurements and Techniques 42
          4.3.1   Instruments 42
          4.3.2   Experimental Setup 45
    4.4   Results of B-field Measurements 47
          4.4.1   Field Mapping 47
          4.4.2   Field Stability Runs 51
          4.4.3   Temperature Dependence 52
          4.4.4   Reproducibility Runs 55
          4.4.5   Summary from Stability and Reproducibility Runs 56
          4.4.6   Current Scans of Magnets 56
          4.4.7   Residuals for Magnets 3B1, 3B2 and 3B4 58
    4.5   Error Sources and Estimations 62
          4.5.1   Error of B-field Integral Monitoring 65
    4.6   Summary 65
    4.7   Recommendations for Future Measurements 66
    4.8   The 4-Magnet Chicane in End Station A 66

5 Relative Beam Energy Resolution   69

    5.1   General Considerations 69
    5.2   Energy BPMs 69
          5.2.1   Energy BPM Resolution 70
    5.3   ESA Magnetic Chicane 73
          5.3.1   Mid-chicane BPM 4 73
          5.3.2   Evaluation of $x_{jitter}^{(4)}$ 74
          5.3.3   Dipole Magnets 76
          5.3.4   Energy Resolution of the Spectrometer 77
          5.3.5   X- and Y-Position Coupling 79
    5.4   Summary 80

6 Laser Compton Energy Spectrometer   83

    6.1   General Considerations 83
    6.2   The Compton Scattering Process 84
          6.2.1   Compton Cross-Section 85
          6.2.2   Properties of the Final State Particles 86

6.3	Overview of the Energy Spectrometer		88
	6.3.1	General Layout	88
	6.3.2	Method A	92
	6.3.3	Method B	95
6.4	Detector Options and Simulation Studies		98
	6.4.1	Photon Detection	98
	6.4.2	Simulation Studies	99
6.5	Laser Power		105
6.6	Potential Background Processes		106
	6.6.1	Multiple Scattering	106
	6.6.2	Nonlinear Effects	110
	6.6.3	Breit-Wheeler Process	113
	6.6.4	Higher Order QED Corrections	113
6.7	Potential Systematic Error Sources		115
	6.7.1	Quartz Fiber Detector	116
	6.7.2	Avalanche Detector	117
	6.7.3	Beam and Laser Jitter	117
	6.7.4	X_γ Determination	117
	6.7.5	X_{edge} Determination	118
	6.7.6	Method A	118
	6.7.7	Method B	119
6.8	Suitable Energy Spectrometer Locations		121
6.9	Summary		123

Conclusions   127

Bibliography   129

Acknowledgments   135

List of Figures

1.1	ILC complex	2
1.2	Electron source	3
1.3	Positron source	4
1.4	ILC RF cavity	4
1.5	Beam Delivery System	6
1.6	Luminosity-weighted $d\mathcal{L}(x)/dx$	8
2.1	Kinematics of the Compton process	13
2.2	Compton spectrum	15
2.3	Particle path within dipole magnet	15
2.4	Generic dipole magnet	16
2.5	Feynman diagram of $e^-e^+ \to Z/\gamma^* \to \mu^+\mu^-\gamma$	17
2.6	Definition of scattering angle	17
2.7	$\sqrt{s}$ reconstruction	18
2.8	Laser Compton energy spectrometer at BESSY I	19
2.9	BESSY II energy measurement example	20
2.10	BESSY II energy measurement example, more details of Fig. 2.9	20
2.11	SLAC energy spectrometer	22
3.1	3-magnet chicane	25
3.2	4-magnet chicane	26
3.3	SLAC	27
3.4	A-line	28
3.5	ESA beam line	29
3.6	Interferometer system	30
3.7	Monopole and dipole modes	32
3.8	Cavity beam coupling	34
3.9	Beam trajectory through a cavity monitor	34
3.10	Mode selection	35
3.11	Polarization superposition	36
3.12	BPM 7 and 9	36
3.13	IQ-plot	38
4.1	Magnets 10D37	42

4.2	Flux gate scheme	44
4.3	Flip coil technique	45
4.4	Test bench table	46
4.5	X-scan measurement	48
4.6	B-field simulation versus X for Z=0	48
4.7	Z-scan	49
4.8	Z-scan simulation comparison	50
4.9	Z-scan simulation comparison, a zoom	50
4.10	Z-scan for zero-current fields, magnet 3B4	51
4.11	Z-scan for zero-current fields, magnet 3B3	51
4.12	Stability runs	53
4.13	Residuals from stability runs	54
4.14	Residual field measurements	54
4.15	Temperature dependence	55
4.16	Reproducibility run, magnet 3B1	56
4.17	Current scan	57
4.18	NMR probe calibration, magnets 3B1 and 3B2	57
4.19	NMR probe calibration, magnet 3B3	58
4.20	Residuals for magnet 3B1	59
4.21	Residuals for magnet 3B2	59
4.22	Residuals for magnet 3B4	60
4.23	Schematic B-field representation	60
4.24	NMR probe and flip coil relative variations, magnet 3B1	61
4.25	NMR, Hall probe and flip coil relative variations, magnet 3B2	62
4.26	Magnetic chicane in ESA	67
5.1	BPM 12 data	71
5.2	BPM 12 normalization	71
5.3	Energy BPM resolution	72
5.4	BPM 4	73
5.5	4-magnet chicane	74
5.6	Evaluation of $x^{(4)}_{jitter}$, linear extrapolation	75
5.7	x4Pos vs. x5Pos and x5Tilt	76
5.8	Nominal and measured B-fields	77
5.9	BPM 9 for an energy scan	77
5.10	Evaluation of $x^{(4)}_{jitter}$, Eq. (5.10)	78
5.11	Beam energy resolution	79
5.12	Resolution of x4Pos - $x^{(4)}_{jitter}$	80
5.13	X- and Y-position coupling	81
6.1	Compton process	84

List of Figures

6.2 Differential and total cross-sections 85
6.3 Energy spectrum dσ/dE and position spectrum dσ/dX of photons 86
6.4 Energy spectrum dσ/dE of electrons 87
6.5 Edge energy versus beam energy 88
6.6 Compton spectrometer 89
6.7 Position spectrum dσ/dX of electrons 91
6.8 Beam energy uncertainty of method A 93
6.9 X_{edge} smearing 94
6.10 Beam energy uncertainty of method B 97
6.11 Diamond strip detector response for X_{edge} determination 101
6.12 Quartz fiber detector response for X_{edge} determination 101
6.13 Position and energy distributions of $e^{\pm}$ particles 103
6.14 Position and energy distributions of surviving SR photons 103
6.15 Quartz fiber detector response for X_{γ} determination 104
6.16 Avalanche detector response 105
6.17 Multiple scattering, extrapolation procedure 108
6.18 Scattered electrons spectrum including multiple scattering 109
6.19 Ratio between unpolarized and polarized Compton cross-sections 110
6.20 Nonlinear QED Feynman diagram 110
6.21 Nonlinear effect, an example 112
6.22 Ratio between first and second harmonics 113
6.23 Breit-Wheeler process, extrapolation procedure 114
6.24 Double Compton and direct pair production 114
6.25 Order-α^3 contributions to the energy spectrum 115
6.26 First and second term ratio of the Taylor series (2.18) vs. beam energy 120
6.27 Possible spectrometer locations 122

List of Tables

1.1	ILC parameters	3
2.1	LEP2 beam energy measurement accuracy	23
3.1	ESA beam parameters	27
4.1	Magnetic center positions	47
4.2	Calibration coefficients of the NMR probe	57
6.1	Contributions to the beam energy error of method B	97
6.2	Simulation parameters	107

Introduction

The largest electron/positron collider built so far was the Large Electron Positron Collider (LEP) [1] at CERN, a storage ring of nearly 27 km of circumference. Here, e^+/e^- particles using the same beam pipe were circulated in opposite direction and accelerated up to the final energy. The machine started operation with a beam energy of ∼45 GeV in 1989 (LEP 1). In 1996, LEP 1 was upgraded and the beam energy was ramped to 81 GeV (LEP 2), with further increases each year up to 104 GeV reached in 2000. After that, the machine was shut down. The advantage of a storage ring compared to a linear collider is that the particles can be kept longer for collisions. Its major drawback is the energy loss per turn due to synchrotron radiation, $U_0 = C_\gamma E_b^4/\rho$, where C_γ is a constant of about $8.86 \cdot 10^{-5} (\text{GeV})^{-3}$, E_b the beam energy and ρ the effective radius of the machine. As can be seen, the energy lost per turn in a storage ring increases dramatically with beam energy. Less energy loss needs an increase of the radius of the machine, but cost problems limit this option. At LEP, the energy lost per turn was compensated by more RF cavities in the straight sections of the collider and higher field gradients created by the klystrons. Further increase of the beam energy beyond the maximum was limited by the available space and the capabilities of the klystrons. At e.g. 100 GeV, the energy loss per turn was 2.9 GeV. At a linear collider, the energy loss during acceleration is negligible, so that for a future collider where electrons and positron are intended to be accelerated beyond 100 GeV beam energy a linear configuration is mandatory.

LEP was built right after the discovery of the Z and W bosons [2, 3] at the Super Proton Synchrotron (SPS) [1] after its conversion to a proton-antiproton collider. The vector bosons Z and W were extensively studied at LEP, and the results obtained achieved an unprecedented precision. Further examples in the past where a hadron machine was used to discover new particles were the discoveries of the bottom and top quarks at Fermilab in 1977, respectively, 1995 [4–6]. The bottom quark was studied in details at lepton colliders, e.g. at LEP and SLC, whereas the top is expected to be studied with best precision at an e^+/e^- linear collider, the (probably) most suitable instrument under clean experimental conditions. Indeed, a hadron machine has an excellent potential to discover new particles within a large mass range. However, the drawback is the presence of large background, mainly due to strong interactions. For that reason it is generally believed that after the discovery of new particles, their properties are better measured with an electron/positron collider. Such a collider suppresses much of the background (due to hadronic interactions) and has the advantage of knowing precisely the parameters of the colliding beams such as energy and polarization, allowing to perform precise measurements.

At present, the Large Hadron Collider (LHC) is proposed for running at CERN. It is a hadron machine built in the tunnel of LEP which accelerates two proton beams. This machine is supposed to answer main questions of the today's particle physics such as the existence of the Higgs boson(s) and of supersymmetric particles, the origin of the dark matter and the existence of extra dimensions. Following LHC, an electron/positron collider is supposed to measure precisely the parameters of possible new particles. Such a facility is under study over the last fifteen years, and is called today the International Linear Collider (ILC). The ILC is a $250 \div 500$ GeV center-of-mass high-luminosity linear collider, based on 1.3 GHz superconducting radio-frequency accelerating cavities. The use of this technology was recommended by the International Technology Recommendation Panel in August 2004 [7] and shortly thereafter endorsed by the International Committee for Future Acceleration [8]. Today, many institutes around the world are involved in linear collider R&D united in a common effort to produce a design for the ILC.

The basic requirements of the ILC are to operate at a center-of-mass energy above $\sqrt{s} = 200$ GeV, upgradeable to 1 TeV, with a design peak luminosity of $2 \cdot 10^{34} \text{cm}^{-2}\text{s}^{-1}$, corresponding to an integrated luminosity of 500 fb^{-1} for the first four years of operation at 500 GeV cms energy. As already mentioned, monitoring the beam parameters is an essential ingredient for precise measurements. At the ILC, energy and polarization are intended to be measured absolutely with a relative accuracy of 0.1‰ and 0.1%, respectively, or better.

Several techniques are proposed to be implemented at the ILC in order to achieve an excellent bunch-to-bunch beam energy control. In particular, energy spectrometers upstream and downstream the of electron/positron interaction point (IP) are believed to be necessary [9]. The default option for the upstream spectrometer is based on a chicane of magnets including beam position monitors (BPMs), the BPM-based spectrometer. In the years 2006/2007, a prototype of such a device was commissioned at the End Station A beam line at the Stanford Linear Accelerator Center in order to study its performance and reliability, denoted as the experiment T474/491 [10–12]. In this experiment, my tasks concerned monitoring the B-field integral of the magnets and to evaluate the energy resolution of the spectrometer.

In addition, a new method for beam energy determination based on laser Compton backscattering was proposed and its feasibility was studied [13]. Here, the average energy of beam particles in a particular bunch is measured, by using the effect that photons from a laser beam can interact with single bunch electrons and from the distinct properties of the scattered particles the beam energy can be deduced. In fact, experiences at LEP and SLC proved that complementary methods of monitoring the beam energy are important and should be implemented in order to cross-check the results of the BPM-based spectrometer.

The thesis is organized as follows. Chapter 1 provides a general description of the ILC accelerator complex including a discussion of measurements which need precise beam energy determination. In Chapter 2 on overview on past experiences on beam energy measurements is given. Chapter 3 describes the goal and layout of the experiment T474/491 in some details. Chapter 4 summarizes the results of the B-field measurements for the magnets of the prototype

BPM-based spectrometer at SLAC. In Chapter 5 the beam energy resolution of the spectrometer is evaluated, while in Chapter 6 the studies performed to evaluate the feasibility of a novel method for fast and precise beam energy monitoring based on laser Compton backscattering are presented. At the end of the thesis, the conclusions are given.

1 The International Linear Collider

In this chapter a general and short description of the current baseline of the ILC as presented in the Reference Design Report (RDR) [9] will be given.

1.1 ILC Basic Design

A schematic view of the ILC complex is given in Fig. 1.1 which indicates the major sub-components:

- a polarized electron source based on a photocathode DC gun;

- an undulator-based positron source, driven by the 150 GeV electron beam, located in the main linac tunnel;

- two damping rings, where the two beams are circulating at 5 GeV;

- beam transport from the damping rings to the main linacs, followed by a two-stage bunch compressor and spin rotator system;

- two 11 km long main linacs;

- a beam delivery system of 4.5 km total length which brings the two beams with a 14 mrad crossing angle to the physics e^+e^- interaction point (IP).

Table 1.1 summarizes major machine and beam parameters.

1.1.1 Electron Source

The electron source is composed by a photocathode in a DC gun illuminated by a laser. Two independent laser and DC gun systems provide redundancy. The main requirements for the electron source imply creation of a bunch train of highly polarized electrons (>80%), to capture and accelerate the beam up to 5 GeV, and to transport the particles to the damping ring. Acceleration of the electrons is done in two steps. In the first step, the bunch is accelerated up to 76 MeV including bunching of the beam. After that, the energy is collimated by means of a vertical 4-magnet chicane and the beam will be accelerated to 5 GeV using superconducting cavities. Prior the damping ring the spin will be rotated to be parallel to the magnetic field in the damping ring and some energy compression is performed. A schematic layout of the polarized electron source is given in Fig. 1.2.

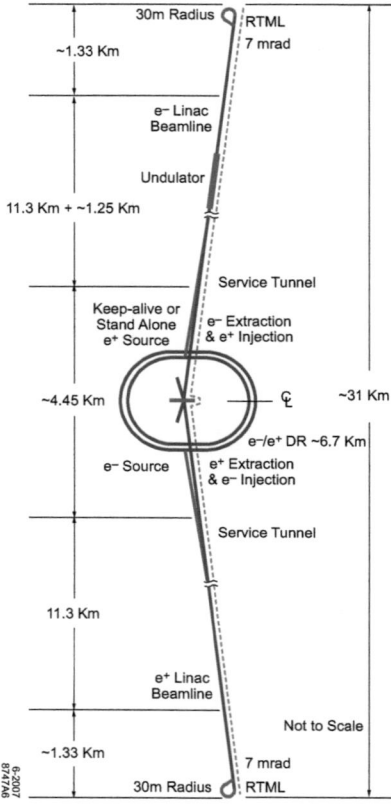

Figure 1.1: Basic layout of the ILC complex at $\sqrt{s} = 500$ GeV. An upgrade to $\sqrt{s} = 1$ TeV requires an extension of the linacs and the beam transport lines by 11 km.

1.1.2 Positron Source

After accelerating to 150 GeV, the electrons pass through a 150 m long helical undulator and afterwards they return to the electron linac. The electrons in the undulator generate high-energy ($\sim$10 MeV) photons which are collimated and directed onto a target about 500 m further downstream. Here, $e^{\pm}$ pairs are produced which are, after some matching and capturing, directed to a first RF cavity and accelerated up to 125 MeV. A dipole magnet selects the positrons and further acceleration to 400 MeV is followed, while the electrons and remaining photons are dumped. Solenoid fields after the $\gamma \rightarrow e^+e^-$ conversion target and in the preacceleration phase reduce the divergence of the positrons so that they are able to match the optic requirements. After that, the particles are accelerated up to 5 GeV using superconducting cavities and then directed to the damping ring. It is expected to generate positrons with a polarization of $\sim 30\%$, with a possible upgrade up to $\sim 60\%$.

1.1 ILC Basic Design

Machine and bunch parameters	Value	Unit
Center of mass energy	500	GeV
Peak luminosity	$2 \cdot 10^{34}$	$cm^{-2}s^{-1}$
Accelerating gradient	31.5	MV/m
Bunch train repetition rate	5	Hz
Bunch train length	1	ms
Number of Bunches per train	2625	
Bunch population	$2 \cdot 10^{10}$	
Linac bunch interval	369	ns
RMS bunch length	300	μm
Typical beam size at IP (horizontal × vertical)	640× 5.7	nm
Normalized emittance at IP (horizontal × vertical)	10×0.04	mm×rad

Table 1.1: Basic design beam and machine parameters for the $\sqrt{s} = 500$ GeV configuration of the ILC.

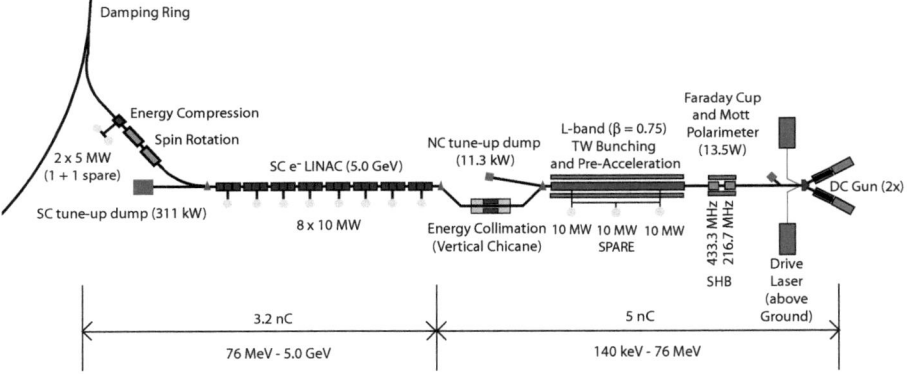

Figure 1.2: Schematic layout of the electron source.

In addition to this source a second option has been proposed. It is called the "Keep-Alive" source (KAS) where the electrons right after preacceleration to ~500 MeV are used to create positrons within a heavy-metal target when the high-energy electron beam is not available. The intensity of the positron beam is, however, lowered to some 10% of the nominal positron beam. Figure 1.3 shows a schematic representation of the positron source.

1.1.3 Damping Rings

At the center of the ILC complex two damping rings exist. Here, the two beams are injected with 5 GeV. The design of the damping rings is constrained by the timing scheme of the linac such that the rings must be large enough to contain a whole train of bunches and, simultaneously, reduce to emittance in less than 200 ms, which corresponds to the train spacing. These rings have a circumference of roughly 6.7 km and are located 10 m above the linac plane to ensure appropriate shielding. The main purpose of the dumping rings is to reduce the horizontal and

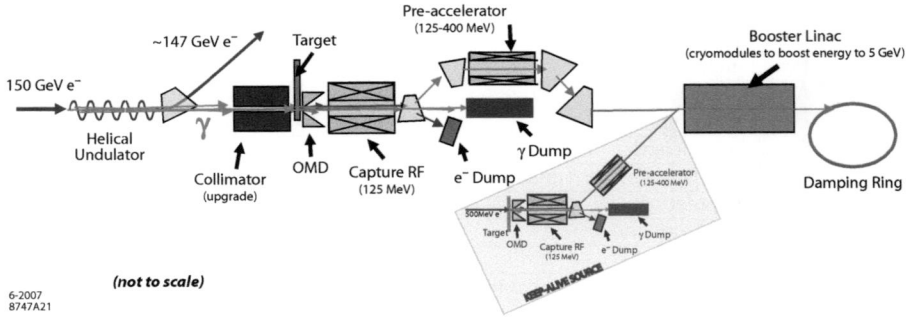

Figure 1.3: Schematic layout of the positron source.

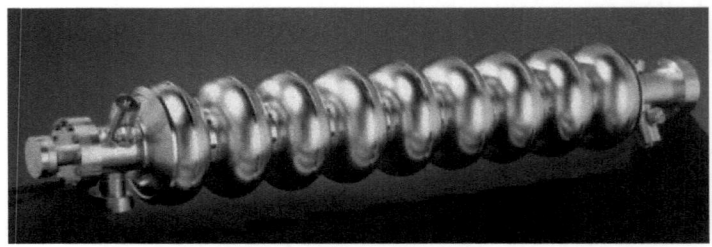

Figure 1.4: A nine-cell 1.3 GHz superconducting niobium cavity.

vertical emittance of the beam through emission of synchrotron radiation. For example, the positron vertical emittance is aimed to be reduced by six orders of magnitude.

1.1.4 Main Linacs

After the damping ring the e^+/e^- beams are transported to the main linacs. Before entering the linacs the particles are accelerated up to 15 GeV and some spin rotation is applied. Also, a 180° turn-around is proposed, which enables feed-forward beam stabilization. The main linac is supposed to accelerate electrons and positron up to 250 GeV, using ~17000 superconducting RF cavities (SCRF cavities). The very low power loss in the SCRF cavity walls allows the use of long RF pulses.

For particle acceleration, 1.3 GHz (L-Band) nine-cell standing-wave niobium cavities (Fig. 1.4) with an average gradient of 31.5 MV/m in 2 K superfluid helium bath are proposed. The supposed gradient is somewhat higher than typical gradients of modern superconducting cavities. The highest gradient obtained so far is 50 MV/m in a single cell cavity, but the most challenging task for the ILC is to be able to bring high gradient cavities to mass-production level. The key objectives for high cavity performance is ultra-clean and defect-free inner surfaces. Hence, preparation and assembly must be made in high-class clean-room environments.

Nine cavities are mounted together to a string and inserted in a common low temperature cryostat, the cryomodule. The total length of a cryomodule is ~12.7 m.

A two-tunnel system (possibly hundreds of meters) underground separated by a distance of $5 \div 7$ m is proposed. The first tunnel, the main tunnel, hosts the acceleration components and the other serves as service tunnel with the RF system, the power supplies and instrumentation racks. Such a schema allows access during beam operation and protects the electronics from radiation damages. Penetrations between the tunnels are foreseen for wave guides, signal and high voltage cables. The main linacs follow the curvature of the earth in order to simplify liquid helium transport. The two linacs are each 11 km long and the upgrade to $\sqrt{s} = 1$ TeV requires an additional extension of about 11 km.

1.1.5 Beam Delivery System

After the main linac the beam delivery system (BDS) follows which transports and focuses the beams to the interaction point (IP). After collision, the spent beams are transported to the main dumps. The BDS is 4500 m long and its main purposes are:

- perform extensive beam diagnostics and match the beams into the final focus;
- protect the beam line and detector against mis-steered beams;
- remove any large beam-halo to minimize the background in the detector.

The layout of the BDS is shown in Fig. 1.5.

Right after the linac, sacrificial collimators are present to protect the beam line in cases of large off-axis beams. High resolution BPMs and kickers provide an intra-train feedback system to correct the trajectory. Four laser wire systems perform emittance diagnostics. They measure the horizontal and vertical beam sizes with a precision of 1 μm. A 4-magnet chicane after the emittance diagnostic system is used for Compton polarimetry and energy diagnostics. In the mid-chicane, a collimator is placed for energy collimation together with a laser-electron beam interaction section, from where backscattered electrons and photons are utilized to measure the beam polarization. An emergency extraction line is placed after this chicane and used to extract the beam in cases of a fault or to dump the beam when not needed at the IP, for example during commissioning of the system.

As seen in Fig. 1.5, an energy collimation system composed of a long chain of dipoles follows. These collimators remove the beam-halo, which can create unwanted background in the detector. In fact, it is required that no particle loss occurs in the last hundreds meter of the BDS and synchrotron radiation must pass cleanly through the IP. Furthermore, a wall shield is placed after the collimation system to suppress muon background generated by electromagnetic showers. The very penetrating muons are shielded from the detectors by this magnetized wall. The magnetic field of the wall has opposite polarities in the right and left halves so that the B-field at the beam line is zero, providing good suppression for the muons and, at the same time, no impact on the orbit of the beam.

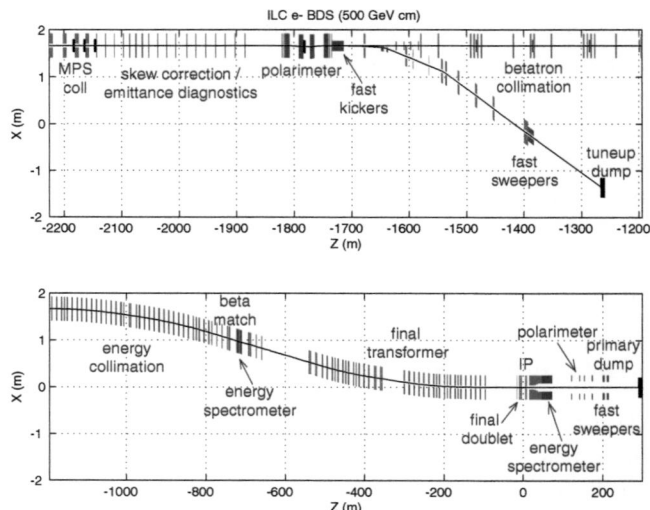

Figure 1.5: Layout of the beam delivery system with the main subsystems starting from the end of the main linac up to the interaction point (IP). The abscissa is the distance from the IP (m) and the ordinate the horizontal position of elements (m).

Beam diagnostics is an important key feature of the beam delivery system. Upstream and downstream of the IP, energy and polarization measurement systems are proposed. For upstream polarization measurement, a Compton polarimeter is employed and, for upstream energy measurement, a BPM-based spectrometer constitutes the default option. Downstream of the IP, a synchrotron radiation spectrometer measures the spent beam energy and a further Compton polarimeter its polarization.

On both sides of the interaction region, calorimeters will be installed very close to the beam pipe to detect particles emitted at small angles. Measuring, for example, the energy deposition from Bhabha events in the angular range $30 \div 90$ mrad, the ILC luminosity is expected to be deduced with a precision of 10^{-3}. At smaller angles, detection of beamstrahlung e^+e^- pairs provides fast monitoring of beam parameters such as the transverse bunch size and the bunch length.

The BDS is supposed to operate, except of few minor changes, also for the $\sqrt{s} = 1$ TeV cms energy upgrade.

For cost saving reasons, a single interaction point shared by two detectors with a "push-pull" option is foreseen at the ILC.

In preparing the Technical Design Report for ILC completed and documented by the end of 2010, some components of the ILC accelerator complex are now redefined toward a more coherent and optimized performance-to-cost-to-risk ratio. Some of the most complex and difficult changes under consideration are a) to replace the double-tunnel configuration by a single-tunnel configuration, b) to redesign the damping ring to smaller circumference and c) to replace the undulator-based positron production source by a conventional source based on Compton backscattering.

1.2 Physics at the ILC

Concerning the physics program, an important task at the ILC is to measure parameters and properties of new particles and couplings. In 1983, the Z and W gauge bosons were discovered at the SPS and 10 years later, LEP an electron-positron collider, performed precise measurements for both particles with a precision in the order of 10^{-4} [14, 15]. Fundamental prerequisites of such measurements were the knowledge of initial beam parameters, for example the energy of the beams. One major advantage of a lepton collider compared to a hadron machine is that definite initial state conditions exist. At the LHC, a proton-proton collider, interactions occur between basic constituents of the hadron, i.e. between quarks and gluons. Here, the energy and momentum carried by these particles vary within relatively broad distributions, described by the parton distribution functions. At a lepton collider, the initial state of the colliding particles is well defined, in particular energy, momentum and polarization are well known and provide important constraints when measuring the properties of new states. An example for such measurements at the ILC is the precise determination of the top quark mass by the so-called "mass scan": counting the number of top-antitop events near the production threshold provides its mass since the cross section rises very fast near $\sqrt{s} = 2m_t$ [9].

At the ILC it is planned to perform such measurements, but some additional drawbacks have to be taken into account. Apart from having sufficiently high luminosity, effects such as beamstrahlung, beam spread and initial state radiation (ISR) modify the original luminosity spectrum $d\mathcal{L}/d\sqrt{s}$ as shown in Fig. 1.6 (right), so that the precise top mass determination is more sophisticated [16].

Tools have been developed to account for such effects. In particular, the acolinearity of Bhabha events provide, together with beam-beam Monte Carlo simulations, a relative differential luminosity spectrum, like the one shown on the right-hand side of Fig. 1.6 [17]. The upper curve represents the luminosity spectrum taking into account all effects, namely the beamstrahlung, the initial state radiation and beam spread. In the lower curve the initial state radiation and spread are counted for, while the sharp curve visible on the right side of the figure shows only the contribution of beam spread. Thereby, the observed $t\bar{t}$ cross section at the nominal center-of-mass energy $\sqrt{s}$ can be written as

$$\sigma_{obs}(\sqrt{s}) = \int_0^1 \frac{d\mathcal{L}(x)}{dx} \sigma(x\sqrt{s}) dx \; , \tag{1.1}$$

where $d\mathcal{L}/dx$ is the differential luminosity, with $x = \sqrt{s'}/\sqrt{s}$ and $\sqrt{s'}$ the effective center-of-mass (cms) energy. As can be seen from Fig. 1.6 (left), the impact of different contributions on the shape of the observed $t\bar{t}$ cross section, $\sigma_{obs}(\sqrt{s})$, is significant and needs careful analysis of the data. In the figure, the upper curve is the cross-section calculated for monochromatic beam

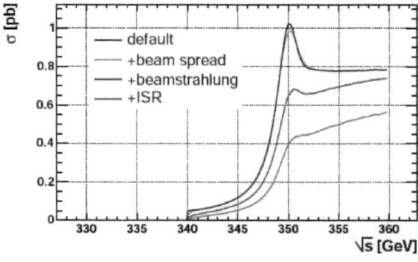 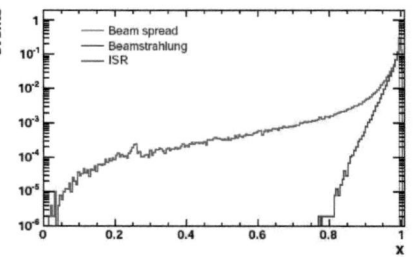

Figure 1.6: Right: Relative luminosity-weighted center-of-mass energy $d\mathcal{L}(x)/dx$. Left: Impact of $d\mathcal{L}(x)/dx$ on the shape of $\sigma_{t\bar{t}}$, where $x = \sqrt{s'}/\sqrt{s}$ and $\sqrt{s}$ the nominal center-of-mass energy, $\sqrt{s} = E_b^{e^+} + E_b^{e^-}$, and $\sqrt{s'}$ the effective cms energy. Default represents the case of a perfect monochromatic beam. The tails at lower energies are due to the particle width.

(indicate as "default" in the figure). The lower curves are the cross-section where additional effects are added up, namely the beam spread, the beamstrahlung and initial state radiation.

In general, the effective center-of-mass energy of the e^+/e^- collision, $\sqrt{s'}$, is $3 \div 4\%$ smaller than the nominal cms energy, $\sqrt{s}$, calculated as the sum of the upstream beam energies of both beams. However, in the analysis for m_t the data are plotted against $\sqrt{s}$ (see left-hand side of Fig. 1.6), whereas in Eq. (1.1) both the effective and nominal cms energy are present. The consequence of this is that the measurement of $\sigma_{t\bar{t}}$ as a function of $\sqrt{s}$ depends on the beam energy error ΔE_b, which is directly connected to the error of the top mass, i.e. $\Delta E_b/E_b = \Delta m_t/m_t$. For physics reasons it is argued to measure the top mass with a relative precision of $3 \cdot 10^{-4}$, which in turn requires to determine the beam energy with an accuracy of 10^{-4}. Otherwise, ΔE_b becomes the main systematic contribution to Δm_t.

Another example of mass measurements at the ILC is the Higgs boson mass determination through the Higgs-strahlung process. Here, the Higgs boson is produced in association with the Z boson. The final states to be taken into account for the analysis are: $HZ \to b\bar{b}q\bar{q}$, $HZ \to b\bar{b}e^+e^-$ and $HZ \to b\bar{b}\mu^+\mu^-$ [18]. It is proposed to apply kinematic fits, imposing four-momentum conservation, to events with a pair of isolated leptons having an invariant mass compatible with the Z mass and 2 jets, or to 4-jet events with a 2-jet invariant mass in accord with the Z boson. The Higgs mass is then reconstructed after the fitting procedure as the invariant mass of the 2 jets assigned to it. Using $\sqrt{s} = 350$ GeV and 500 fb^{-1} accumulated luminosity, simulations indicate that a statistical error of 40 MeV for the Higgs mass m_H can be achieved. Dedicated studies were performed to investigate the impact from the error of the beam energy, the beam spread and uncertainties in the differential luminosity spectrum on m_H [19]. It was found that the systematic error on the Higgs mass depends linearly on the uncertainty of the beam energy as

- $\Delta m_H \sim 0.8 \cdot \Delta E_b$ for the $HZ \to b\bar{b}q\bar{q}$ channel and

- $\Delta m_H \sim \Delta E_b$ for the $HZ \to b\bar{b}l^+l^-$ channel.

To keep this systematical error below the statistical one, the beam energy has to be measured with a precision of 10^{-4}. In addition, the beam spread increases the statistical error of m_H by additional $5 \div 10$ MeV and the parameters of the luminosity parametrization must be known with an uncertainty of 1% in order to keep Δm_H below $40 \div 50$ MeV.

These two examples of mass measurements imply precise evaluation of E_b. A further argument which points to perform beam energy measurement with high precision is the estimation of the integrated luminosity, $\int \mathcal{L} dt$, which is used to calculate any reaction cross-section from the recorded number of events. The technique used to determine $\int \mathcal{L} dt$ is based on counting Bhabha events since the cross-section for this process is a priori very well known. The cross-section is proportional to the inverse of the square of the center-of-mass energy, hence, a wrong E_b estimation will lead to a mismeasurement of the luminosity.

Machine simulations revealed that E_b might have a large jitter between bunches and, in addition, a strong head-tail effect inside the train might be present. These effects were estimated to be in the same order of magnitude as the beam energy spread of $\sim 10^{-3}$. Thereby, monitoring the beam energy bunch-to-bunch with a precision of 10^{-4} is necessary to reconstruct the distribution of E_b within the train. This information can also be used as additional input for an improved parametrization of the differential luminosity $\mathcal{L}(x)$. Also, bunch-to-bunch beam energy measurements have some fundamental importance for the study of possible correlations between electron and positron beam bunches.

2 Beam Energy Measurement Techniques

In this chapter a review on most common techniques for beam energy measurements will be presented. In Sect. 2.1 some details on several well established methods are outlined. Section 2.2 discusses applications of these methods in the past and the results obtained at electron-positron linear colliders and storage rings where accurate beam energy determinations were requested. The last section, Sect. 2.3, describes suitable options for beam energy measurements at the ILC.

2.1 Review on Methods

2.1.1 Resonant Depolarization

In a storage ring, electrons and positrons become transversely polarized through the Sokolov-Ternov effect [20] due to the emission of synchrotron radiation. The time-dependent level of the polarization is given by

$$P_e = P_0(t - e^{-t/\tau_0}) \, . \tag{2.1}$$

The time constant τ_0 depends on the energy and geometry of the storage ring. For LEP at E_b=45 GeV for example, τ_0 was 5.7 hours. The upper limit of the polarization, P_0, is 92.4%. The precession of the spin $\vec{s}$ of a relativistic particle is described by the so-called Bargmann-Michele-Telegedi equation [21]

$$d\vec{s}/dt = \vec{s} \times \vec{\omega}_{BMT} \, , \tag{2.2}$$

with

$$\vec{\omega}_{BMT} = \frac{c}{E_b}\left[(\gamma a + 1)\vec{B} - a\frac{\gamma^2}{\gamma+1}(\vec{\beta}\cdot\vec{B})\vec{\beta} - \left(\gamma a + \frac{\gamma}{\gamma+1}\right)(\vec{\beta}\times\vec{E})\right] \, . \tag{2.3}$$

E_b is the particle energy, γ the Lorentz factor, $\vec{\beta} = \vec{v}/c$, a the anomalous magnetic moment of the electron and $\vec{B}$ and $\vec{E}$ are the magnetic and electric fields, respectively. If we assume $\vec{E} = 0$ and $\vec{B}_{\parallel} = 0$, with $\vec{B}_{\parallel}$ as the fraction of the magnetic field along the beam direction, the spin precesses around the vertical Y-axis with a frequency $\omega_{BMT} = \omega_c(1+\gamma a)$, where $\omega_c = eB_y/m_e\gamma$ is the rotation frequency of the beam in the storage ring. It is customary to define the spin tune [22]

$$\nu_s = \gamma a = \frac{\omega_{BMT}}{\omega_c} - 1 \ . \qquad (2.4)$$

Resonant depolarization is produced by exciting the beam with an oscillating magnetic field generated by a vertical kicker magnet. This field is perpendicular to the beam axis and the bending field of the ring. If the frequency of this horizontal field is in phase with the spin precession and the revolution frequency ω_c, a resonance condition occurs and polarization disappears [23]. More generally, depolarization occurs if a weak depolarizing field with frequency ω_D is applied and the following condition is satisfied [22]:

$$n = n_s \nu_s \pm n_x Q_x \pm n_y Q_y \pm n_z Q_z \pm \left(\frac{\omega_D}{\omega_c}\right) \ . \qquad (2.5)$$

Here, n is an integer, Q_x and Q_y are the betatron and Q_z the synchrotron tunes. At the lowest mode with $n_s = 1, n_x = n_y = n_z = 0$, this relation can be rewritten as [23]

$$\omega_D = (k \pm [\nu_s]) \cdot \omega_c \ , \qquad (2.6)$$

with k an integer and $[\nu_s]$ the fractional part of ν_s. Its integer part is determined from the setting of the bending field.

According to Eq. (2.4), the spin tune and the beam energy are related

$$\nu_s = \gamma a = \frac{aE_b}{mc^2} \ , \qquad (2.7)$$

so that for an electron-positron storage ring (like LEP) the spin tune is

$$\nu_s = \frac{E_b[\text{MeV}]}{440.6486(1)[\text{MeV}]} \ . \qquad (2.8)$$

If ν_s is measured, the beam energy E_b can be determined with very high accuracy. To measure ν_s, an oscillating RF B-field is applied after the beam is transversely polarized. The frequency of this field, ω_D, is swept within a certain interval and, if necessary, repeated for different intervals until depolarization is observed.

The accuracy of the resonant depolarization method is exceptional and if implemented in storage rings an absolute beam energy error up to $2 \cdot 10^{-6}$ has been achieved. The drawback of this option of E_b determination is the existence of a transversely polarized beam and the time to polarize the beam through the Sokolov-Ternov effect. This implies that this method cannot be used at a linear collider. Even in storage rings there are limitations. For example at LEP, the method could not be applied for beam energies above 61 GeV, since it was not possible to generate transversely polarized beams [24, 25].

2.1.2 Compton Backscattering

Compton scattering is a well-known process where a photon scatters elastically with an electron [26]

$$e + \gamma \to e + \gamma \ . \tag{2.9}$$

When a photon with an energy between 100 keV and 10 MeV travels through matter, it can interact with an electron which is bounded to an atom. Since the energy of the electron is very small compared to that of the photon, it can be considered at rest and free. So, in the rest frame of the electron (Fig 2.1) the final state energy of the photon is given by

$$E_\gamma = \frac{E_\lambda}{1 + (1 + \cos\theta_\gamma) E_\lambda/m} \ , \tag{2.10}$$

with $\pi - \theta_\gamma$ as the scattering angle, E_λ the initial energy of the photon and m the rest mass of the electron.

For $\theta_\gamma = 0$, the energy of the scattered photon is smallest, while that of the electron is largest and is called the Compton edge energy.

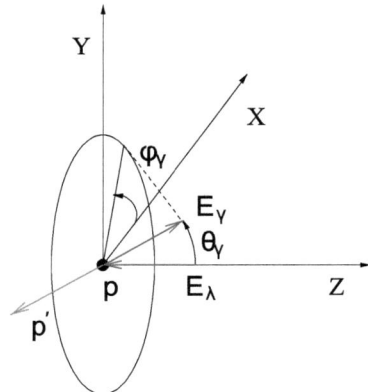

Figure 2.1: Kinematics of Compton scattering in the rest frame of the initial electron.

If the initial electron is not at rest but moves along the Z-direction with some energy as in a collider, the energy of the scattered electron/photon is simply obtained by a boost γ from the electron rest to the laboratory frame. Considering an initial photon which moves along the Z-direction and collides head-on with the electron, the energy of the scattered photon in the laboratory system results in

$$E_\gamma^{Lab} = \gamma E_\gamma (1 + \beta \cos\theta_\gamma) \ , \tag{2.11}$$

with $\gamma = E_b/m$ and $\beta = v/c$; v is the electron velocity and c the speed of light.

Moreover, E_λ, the initial energy of the photon in the electron rest frame, can be expressed by the initial photon energy in the lab frame E_λ^{Lab}

$$E_\lambda = \gamma E_\lambda^{Lab}(1+\beta) \ . \qquad (2.12)$$

Finally, according to Eqs. (2.10), (2.11) and (2.12), the energy of the scattered photon in the lab frame results in

$$E_\gamma^{Lab} = E_\lambda^{Lab} \frac{1+\beta \cos\theta_\gamma}{1-\beta + (1+\cos\theta_\gamma)E_\lambda^{Lab}/E_b} \ , \qquad (2.13)$$

and the relation between θ_γ and θ_γ^{Lab}, the scattering angle in the lab system, is

$$\tan\theta_\gamma^{Lab} = \frac{\sin\theta_\gamma}{\gamma(\cos\theta_\gamma + \beta)} \ . \qquad (2.14)$$

It is interesting to note that the maximum energy of the scattered photon in the lab frame corresponds to the smallest energy of the photon in the rest frame. Indeed, the maximum energy of the scattered photon in Eq. (2.13) is obtained for $\theta_\gamma = 0$

$$E_{\gamma,max}^{Lab} = E_\lambda^{Lab} \frac{1+\beta}{1-\beta + 2E_\lambda^{Lab}/E_b} \ . \qquad (2.15)$$

As can be seen, the boost suppresses the photon scattering angle in the lab frame by γ. In general, an electron beam colliding head-on with laser light results in photons which are 'back' scattered and concentrated in a very small cone of aperture θ_γ^{Lab}. This is the reason to call the process $e + \gamma \to e + \gamma$ as Compton backscattering (CBS). In particular, backscattered photons with maximum energy, $E_{\gamma,max}^{Lab}$, have $\theta_\gamma^{Lab} = 0$.

From now on we will, if not otherwise specified, refer to quantities defined in the lab system. If the electron beam is not polarized, the energy spectrum of the scattered photons is given by

$$\frac{d\sigma_c}{dy} = \frac{2\sigma_0}{x}\left[\frac{1}{1-y} + 1 - y - 4r(1-r)\right] \ , \qquad (2.16)$$

where $y = E_\gamma/E_b$, $x = 4E_b E_\lambda/m^2$ and $r = y/[(1-y)x]$. Equation (2.16) diverges for $y \to 1$, that is for $E_\gamma \to E_b$, but kinematic constraints impose a limit on E_γ, being in the range $E_\gamma \in [E_\lambda, E_{\gamma,max}]$. The energy of the incoming laser photon, E_λ, is usually very small, especially if compared with typical beam energies and is therefore often considered to be zero.

A typical photon spectrum of backscattered photons is shown in Fig. 2.2. As can be seen, the spectrum has a cut-off at $E_{\gamma,max}$, the Compton edge. This maximum energy, if precisely measured, provides access to the beam energy E_b via Eq. (2.15). In the past, beam energy measurements based on Compton backscattering relied on the determination of $E_{\gamma,max}$.

2.1.3 Deflection in a Dipole Field

When a particle passes through a perfect homogeneous magnetic field, it is bent by the Lorentz force in the plane perpendicular to the field direction, with a circular trajectory of radius $R[\text{m}] = p_\perp[\text{GeV}]/(K_b \cdot B[\text{Tesla}])$, where the constant K_b is 0.299792458. If the particle is

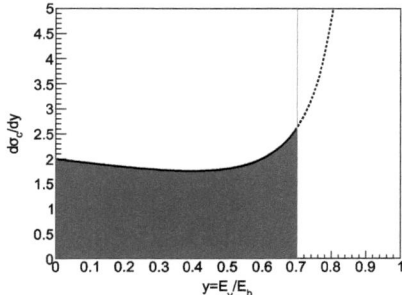

Figure 2.2: Photon spectrum $d\sigma_c/dy$ as a function of the backscattered photon energy $y = E_\gamma/E_b$. The vertical line shows the cut-off energy $E_{\gamma,max}$, while $d\sigma_c/dy$ diverges for $E_\gamma/E_b \to 1$ as indicated by the dashed line.

relativistic and the momentum perpendicular to the magnetic field, we can replace the modulus of the momentum by its energy E_b.

Figure 2.3 illustrates the path of a charged particle entering perpendicularly a dipole magnet of length l. From the figure one derives for the displacement of the particle, X, downstream of the magnet

$$X = R - \frac{l}{\tan\theta} + L \cdot \tan\theta \;, \tag{2.17}$$

with $\tan\theta = l/\sqrt{R^2 - l^2}$. Expanding Eq. (2.17) as a function of l/R one obtains

$$X = \left(L + \frac{l}{2}\right) \cdot \frac{l}{R} + \frac{L}{2} \cdot \left(\frac{l}{R}\right)^3 + \mathcal{O}\left(\left(\frac{l}{R}\right)^5\right) \;. \tag{2.18}$$

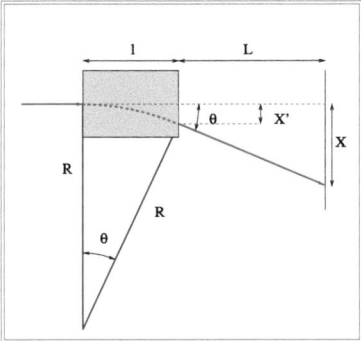

Figure 2.3: Path of a charged particle inside a magnet with a field perpendicular to the particle momentum.

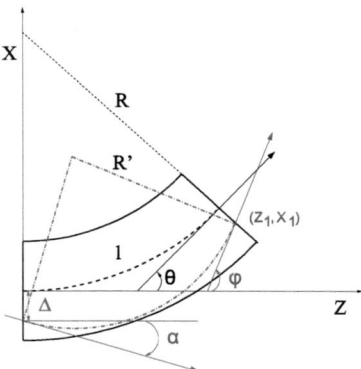

Figure 2.4: Schematic representation of a particle trajectory through a sector dipole magnet with homogeneous field and sharp edge field with the nominal trajectory indicated by the dashed line. The particle enters the magnet with an angle α and an offset Δ. During traversing the magnet the particle is represented by the point-dashed line and at the exit its position is (Z_1, X_1).

For high energy electrons, l/R is often very small, so that Eq. (2.18) can be rewritten as

$$X \simeq \left(L + \frac{l}{2}\right) \cdot \frac{l}{R} = \left(L + \frac{l}{2}\right) \cdot \frac{K_b B l}{E_b} . \tag{2.19}$$

When the beam enters the magnet not perpendicularly but with a small angle α and if $\alpha \ll \theta$, the exit angle of the particle is simply $\theta' = \theta + \alpha$.

Equation (2.19) reveals that a magnet introduces a coupling between the coordinate X and the beam energy, so that measuring the displacement X somewhere downstream of the magnet provides a measurement of the beam energy E_b, provided the product Bl or the B-field integral is known. In a more general way, the formula for the position of a charged particle at the exit of a generic magnet with B-field B, length l and radius R (see Fig. 2.4) becomes

$$\begin{cases} Z_1 = \dfrac{A \cdot R'}{\tan\left(\phi + \frac{\pi}{2}\right) - \tan\left(\frac{\pi}{2} + \theta\right)} \\ X_1 = \tan\left(\frac{\pi}{2} + \theta\right) Z_1 + R \end{cases} ,$$

with R' the radius of the particle trajectory inside the magnet, α the entrance angle, Δ the offset from the magnet center at the entrance in the Z-X plane and

$$\tan\left(\phi + \frac{\pi}{2}\right) = \frac{\tan\left(\frac{\pi}{2} + \theta\right) + A\sqrt{\tan\left(\frac{\pi}{2} + \theta\right) + 1 - A^2}}{1 - A^2}$$

$$\text{and } A = \frac{R - \Delta - R'\cos\alpha + R'\sin\alpha \tan\left(\frac{\pi}{2} + \theta\right)}{R'} .$$

2.1.4 Radiative Return Events

Considering the process

$$e^-e^+ \rightarrow Z/\gamma^* \rightarrow \mu^+\mu^-\gamma \qquad (2.20)$$

with the corresponding Feynman diagram shown in Fig. 2.5.

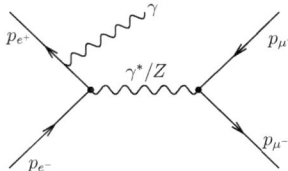

Figure 2.5: Feynman diagram for the process $e^-e^+ \rightarrow Z/\gamma^* \rightarrow \mu^+\mu^-\gamma$.

At sufficiently high $\sqrt{s}$, the incident electron and positron annihilate into a γ^* or the Z boson which then decays into a lepton pair, in our example into muons. One of the initial state particles emits from time to time a photon along its direction (initial state radiation, ISR). If one knows the angle θ_1 and θ_2 of the two final state leptons with the emitted photon (see Fig. 2.6) as well as their invariant mass $\sqrt{s'}$, it is possible to infer the invariant mass or center-of-mass energy $\sqrt{s}$ of the initial state electron/positron pair from

$$\sqrt{s} = \frac{\sqrt{s'}}{\sqrt{1-\kappa_\gamma}}, \qquad (2.21)$$

where

$$\kappa_\gamma = 2\frac{\sin(\theta_1 + \theta_2)}{\sin\theta_1 + \sin\theta_2 + \sin(\theta_1 + \theta_2)}. \qquad (2.22)$$

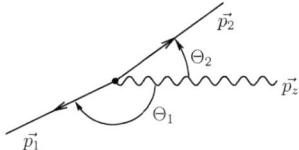

Figure 2.6: Scattering angles θ_1 and θ_2 of the two muons with the emitted photon. If the photon is generated along the Z-direction, these angles can be considered as the scattering angles in the lab frame.

Muons in the final state are highly preferable for such a study because they can be very well identified and accurately measured. Since the photon is emitted at very small angle, θ_1 and θ_2 can be considered as the muon scattering angles in the cms or lab frame. Moreover, considering only events with a di-muon invariant mass consistent with the Z boson mass m_Z, e.g. between 86 and 96 GeV, $\sqrt{s'}$ in Eq. (2.21) can be substituted by the Z mass and the

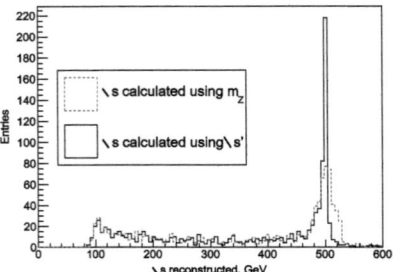

Figure 2.7: Reconstructed center-of-mass energy $\sqrt{s}$ using Eq. (2.21) (full line) and Eq. (2.23) (dashed line).

nominal center-of-mass energy is calculated from the two scattering angles θ_1 and θ_2 [27]

$$\sqrt{s} = \frac{m_z}{\sqrt{1-\kappa_\gamma}}. \tag{2.23}$$

Figure 2.7 shows the reconstructed center-of-mass energy of Eq. (2.21) (full line) and of Eq. (2.23) (dashed line). Each distribution shows a narrow and pronounced peak at the correct energy value. In both cases, the simulation input for $\sqrt{s}$ was set to 500 GeV. Since m_Z has been used instead of $\sqrt{s'}$, the peak in the dashed distribution at 500 GeV is more spread out and events with $\sqrt{s}$ bigger than 500 GeV are more abundant. The tails observed in either case are due to events where at least one of our assumptions failed.

Using the approach of radiative return events for $\sqrt{s}$ determination, accumulation of many $e^-e^+ \to \mu^+\mu^-\gamma$ events is necessary which is a time consuming process (several months) before a precise $\sqrt{s}$ value can be deduced. With a integrated luminosity of 100 fb^{-1}, a relative error of $\sim 1.3 \cdot 10^{-4}$ might be obtained [27].

2.2 Energy Measurements in the Past

2.2.1 BESSY I and II

At the electron storage rings BESSY I and II in Berlin, fast beam energy monitoring based on Compton backscattering has been performed independently on the resonant depolarization method. Both storage rings are light sources used by, amongst others, the Physikalisch-Technische Bundesanstalt(PTB) [28, 29], the German national metrology institute entrusted with the realization and dissemination of the legal units. For such a task the photon flux in the storage rings has to be known with high precision. One fundamental input parameter for the photon flux determination constitutes the beam energy. For BESSY I, which was operating with a beam energy between 340 MeV and 900 MeV, an accuracy of 10^{-4} was required, whereas for BESSY II, operating between 900 MeV and 1.7 GeV, a beam energy precision of better

2.2 Energy Measurements in the Past

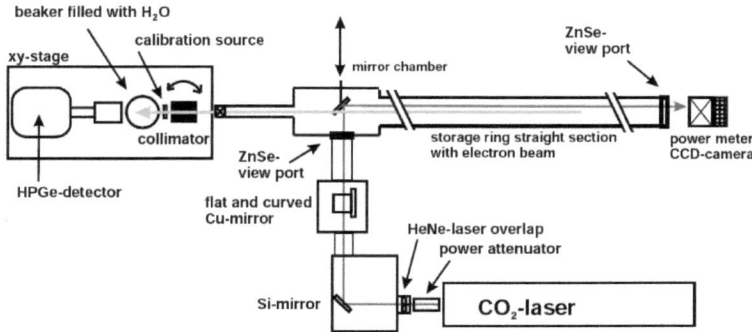

Figure 2.8: Layout of the energy spectrometer based on Compton backscattering at BESSY I and II.

than $5 \cdot 10^{-5}$ was demanded to evaluate the photon flux for photon up to 50 keV with a relative uncertainty below 0.2%. Employing the resonant depolarization method, such an accuracy has been achieved in both machines. However, at energies below 340 MeV at BESSY I and 900 MeV at BESSY II, the time to polarize the beam through the Sokolov-Ternov effect was too long. Hence, another method was searched for and the Compton backscattering turned out to be a suitable option. Figure 2.8 shows the layout of the experiment as performed at both storage rings.

Light from a continuous CO_2 laser was directed to a straight section of the storage ring by an optical system. Here, the electrons interact with the laser beam and produce Compton backscattered photons, which were detected by a high-purity Germanium (HPGe) detector. Since the rate of the backscattered photons was relatively low, less than one for a laser-electron bunch crossing, an energy determination of single photons was possible and, hence, the reconstruction of their energy spectrum. To increase the signal-to-noise ratio and to allow for a not too high counting rate, a collimator was placed in front of the Germanium detector. Since for beam energy measurements only those photons with an energy close to the maximum value, $E_{\gamma,max}$, and a scattering angle θ_γ close to zero are suitable, the collimator removes only unwanted Compton photons and background and did not affect the energy measurement. Calibration of the detector was performed using radioactive sources. At BESSY I, since $E_{\gamma,max}$ was around 1.1 MeV for 800 MeV beam energy, a ^{60}Co source was suitable, since it emits photons with energy of 1.17 and 1.3 MeV and was placed before the collimator. For BESSY II, $E_{\gamma,max}$ is around 5 MeV for a beam energy of 1.7 GeV, so a ^{244}Cm/^{13}C source with photons of $\sim$ 6 MeV was more appropriate. The typical time scale for recording beam energy data at both machines was 15 minutes. Figures 2.9 and 2.10 (left) display the results for an energy monitoring run at BESSY II.

Figure 2.9 shows only the high energy part of the whole spectrum recorded. The Compton edge and the two calibration lines are clearly visible. Their zoom is shown in Fig. 2.10 (left). The Compton edge reveals a step behavior as expected (see Fig. 2.2), but convoluted with the

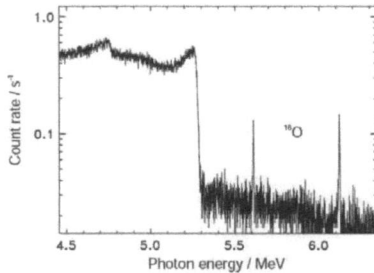

Figure 2.9: The photon spectrum using the Compton backscattering technique. Calibration photons were recorded simultaneously with Compton photons

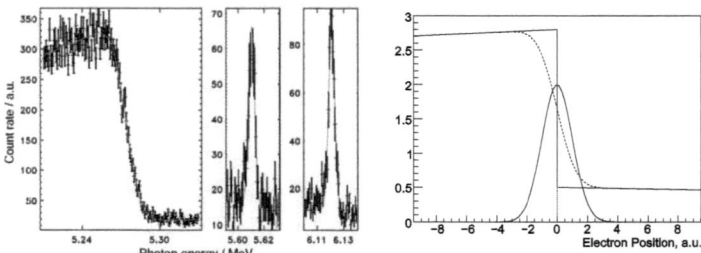

Figure 2.10: Left: A zoom of Fig. 2.9 to show the Compton edge and two calibration lines in more details. Right: The Compton edge approximated by a convolution of an error function (dotted line) and a Gaussian (continuous line).

energy spread of the beam and the detector resolution. Fitting the data by an error function convoluted with a Gaussian, the value of the Compton edge, or $E_{\gamma,max}$, is given by the peak value of the Gaussian as indicated in Fig. 2.10 (right).

As already mentioned, besides the beam energy inferred from Compton backscattering the energy was also independently measured by the resonant depolarization method. Very good agreement, in particularly for BESSY II, was found within 70 keV between both measurements.

2.2.2 VEPP-4M

The VEPP-4M machine is an electron-positron collider at the Budker Institute of Nuclear Physics (BINP) in Novosibirsk. It delivers a maximum beam energy of 6 GeV. VEPP-4M provided the world's highest precision measurement on the τ lepton mass [30]: the τ mass was measured with an error of 0.15 MeV by an energy scan, where the error of the beam energy contributed approximately 40 keV. Running the machine between 1.7 and 1.9 GeV, this corresponds to a relative beam energy precision of $2 \cdot 10^{-5}$.

As at BESSY, resonant depolarization as well as Compton backscattering methods were performed at VEPP-4M [31]. To perform resonant depolarization measurements, the machine was needed to run in a special mode in which two bunches were injected, one polarized and the

2.2 Energy Measurements in the Past

other not. Such an operation reduced the systematic errors substantially. After two hours of measuring, an absolute beam energy accuracy of 2 keV was achieved.

Since rather large energy variations between calibration runs were observed, the beam energy was independently monitored by guide field measurements assuming that $E_b = \alpha_H \cdot H$, where H is the B-field of a reference magnet and α_H a constant. Temperature and mechanical variations required corrections to the formula and were taken into account.

As a third option for monitoring the beam energy Compton backscattering was employed. The setup at VEPP-4M was similar to that at BESSY with one exception: instead of a collimator, absorbers were utilized to optimize the signal-to-noise ratio as well as the counting rate. Due to a coupling between beam energy and position of the photons in X-direction, a misplaced collimator as used at BESSY could introduce some additional systematic error. For beam energies between 1.7 and 1.9 GeV, $E_{\gamma,max}$ was about 6 MeV which allowed to utilize polyethylene as absorber material before the HPGe detector. For detector calibration purposes only γ-sources with energies between 0.5 and 2.7 MeV were available, which required an extrapolation to 6 MeV, the expected Compton edge energy.

Typical data taking periods took $10 \div 60$ min. A beam energy accuracy of typically $40 \div 50$ keV was obtained and the energies measured were in good agreement with the results from the other two methods applied.

2.2.3 Stanford Linear Collider

The Stanford Linear Collider (SLC), a two miles electron/positron linear collider, accelerated the particles up to a center-of-mass energy of ~91 GeV, the mass of the Z boson. In order to perform its physics objectives, an accurate monitoring of the beam energy with a relative error of less than $5 \cdot 10^{-4}$ was demanded. For this task an energy spectrometer after the e^+/e^- interaction point (IP) within the extraction line was proposed [32]. The spectrometer involves three magnets (Fig. 2.11) and one of them, the spectrometer magnet, was well field-mapped and continuously monitored for $\int Bdl$ information. The beam displacement X, which equals to the distance between the two horizontal synchrotron swaths, provides, together with the field integral and the distance L, the beam energy, see Eq. (2.19). The swaths were produced by two ancillary magnets, one placed before and the other behind the spectrometer magnet.

The field integral $\int Bdl$ was measured by means of the moving wire and the flip coil techniques and in situ continuously monitored by the flip coil and probes [33]. Details concerning techniques for local B-field and $\int Bdl$ measurements will be discussed in Chapter 4.

The two synchrotron radiation stripes (or swaths) were measured either by two phosphorescent screens, both were overlayed by an array of fiducial wires [34], or by a wire imaging synchrotron radiation detector (WISRD) [35]. The distance of the stripes was determined with a precision of better than 0.02%. The beam energy was measured bunch-to-bunch with a resolution of 5 MeV and, together with a total systematic error of 20 MeV at $E_b \simeq 50$ GeV, a relative beam energy uncertainty of $3 \div 4 \cdot 10^{-4}$ was achieved.

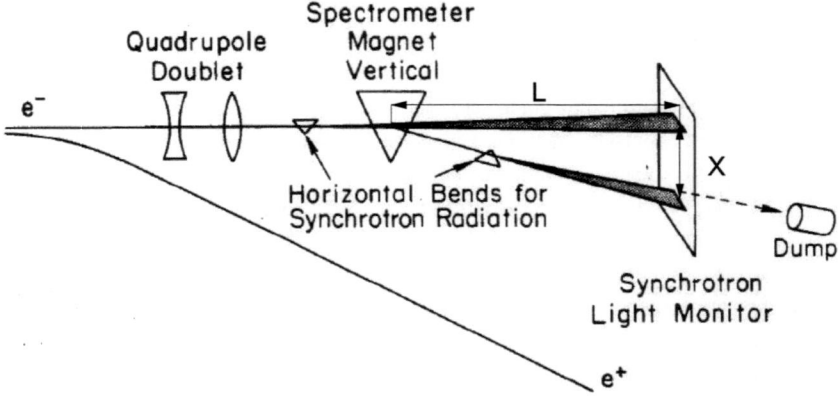

Figure 2.11: Schematic representation of the energy spectrometer at SLAC.

2.2.4 Large Electron Positron Collider

The Large Electron-Positron Collider (LEP) was a storage ring to collide electrons and positrons accelerated up to about 100 GeV. The program started in 1989 with LEP 1, with a maximum energy of about 45 GeV. In 1996-2000, the machine was successively upgraded to LEP 2 with a beam energy up to 104 GeV. Among many measurements performed at LEP 1 (and at SLC), the Z boson mass and its width were measured with unprecedented accuracy, whereas at LEP 2 the W boson mass was measured with an uncertainty of $30 \div 40$ MeV. In both cases, precise beam energy measurements were prerequisites to achieve such outstanding results. For example, the beam energy error translates directly into the error of the W mass, i.e. $\Delta W/W = \Delta E_b/E_b$, in a similar manner as for the Higgs mass proposed to be measured at the ILC. In particular, several techniques to deduce the beam energy were implemented at LEP 2, the resonant depolarization (RDP) method, the NMR model, the flux loop method, a BPM-based spectrometer and the method of synchrotron radiation tune [25]. Resonant depolarization was also the only technique which provided absolute energy determination, whereas all other techniques needed calibration.

Among all these methods, RDP provided the best accuracy, being about few MeV for beam energies in the range of $41 \div 61$ GeV. The method for continuous monitoring, however, was the NMR model, which is based on B-field measurements in few dipole magnets inside the ring. The relation between these measurements and the beam energy was assumed to be linear and, after calibration using the resonant depolarization technique and taking into account possible corrections to the linear relation, fast monitoring of E_b was possible. As already mentioned (see Sect. 2.1.1), due to strong depolarization effects transversely polarized beams with energies above 61 GeV were not available. This implies that resonant depolarization calibration was only feasible for $E_b \lesssim 61$ GeV and could not be performed for runs in the LEP 2 physics regime $81 < E_b < 104$ GeV. Here, only the NMR model was applicable at the beginning.

For redundancy and cross-checks, further techniques were developed and commissioned in the LEP tunnel. The first method, already utilized at LEP 1, was the flux loop method. Throughout many of the bending magnets of the machine a loop was embedded to the lower pole, which allowed to determine the uncertainty of E_b to 7.5 MeV at 72 GeV and to 17.6 MeV for 106 GeV. Moreover, it was found that the difference between the flux loop and the NMR model, $E_b^{FL} - E_b^{NMR}$, was 6 MeV at 100 GeV beam energy.

The second approach relies on a BPM-based spectrometer. Using the idea as described in Sect. 2.1.3, a magnet, the spectrometer magnet (being part of the lattice of LEP), was supplemented by two stations of three beam position monitors, placed before and after the magnet. The spectrometer magnet was accurately mapped and four NMR probes were permanently positioned inside to monitor the B-field in situ during runs [36]. The spectrometer, calibrated at low energies (41 ÷ 61 GeV) using the RDP method, delivered beam energy values E^{SP}, which were compared at higher energies with those from the NMR model. An offset and uncertainty $E_b^{SP} - E_b^{NMR}$ of -4.9 ± 17.8 MeV was found at 92 GeV, whereas at 70 GeV the offset was -0.6 ± 9.7 MeV.

Last but not least, E_b was also measured using the synchrotron tune Q_s. This quantity is related to the beam energy once other informations such as the energy loss from synchrotron radiation and the RF voltage are known. It was found that in the physics regime of LEP 2 the difference $E_b^{Q_s} - E_b^{NMR}$ results in -2.8±15.8 MeV.

Using all the data available, the total error of the center-of-mass energy $\sqrt{s}$ could be estimated. It turned out to be not a constant but varies with $\sqrt{s}$. Table 2.1 summarizes the total uncertainties obtained as a function of the center-of-mass energy.

$\sqrt{s}$ [GeV]								
161	172	183	189	192	196	200	200	205	207
Error [MeV]								
25.4	27.4	20.3	21.6	21.6	23.2	23.7	23.7	36.9	41.7

Table 2.1: Errors on $\sqrt{s}$ as a function of the nominal center-of-mass energy.

2.3 Beam Energy Measurement at the ILC

In the previous sections some relevant and successful methods for beam energy measurements were discussed including special aspects of e^+/e^- storage rings, respectively, linear colliders. The most precise method found so far is the resonant depolarization method, which provides a relative beam energy uncertainty of few ppm. This method, however, is precluded to be applied at a linear collider such as the ILC.

Radiative return events also deliver a well established absolute E_b measurement. Unfortunately, this method requires sufficient integrated luminosity of typically 100 fb^{-1} in order to achieve a relative error of $\sqrt{s}$ of $\sim 1.3 \cdot 10^{-4}$. Therefore, it cannot be used for fast beam energy monitoring but provides valuable checks of the collision energy scale, in particular for long-term calibration of beam energy measurements.

At the ILC, the baseline method to measure E_b is the BPM-based spectrometer. This method is suitable for a very large beam energy range. However, at too low energies limited precision on B-field measurements[1], while at very high energies possible beam emittance growth may preclude E_b determination. A spectrometer as implemented in a storage ring such as LEP, or in linear colliders as the SLC, provided valuable bunch-to-bunch beam energy measurements. Calibration of the device is an issue and has to be accounted for right from the beginning.

In general and based on the physics objectives at the ILC, evaluation of the luminosity-weighted center-of-mass energy requires the knowledge of the beam energy upstream of the physics e^+/e^- IP, see Sect 1.2. Concerning the resolution, a bunch-related value of $\Delta E_b/E_b \simeq 10^{-4}$ is required to ensure that the beam energy error does not become the main systematic error for mass determination.

Taking all arguments together, the ILC design report proposes to install upstream of the e^+/e^- IP a BPM-based spectrometer, which is best located after the energy collimation section and before the muon wall. Since redundancy for beam energy measurement is very important as was demonstrated at LEP, an additional spectrometer is proposed to be placed in the beam extraction line, the downstream spectrometer. It is very similar to the spectrometer used at SLC, but the WISRD detector is replaced by quartz fibers.

However, when the beams are in collisions, the energy measured by the downstream spectrometer is different from the upstream one. Therefore, it is worthwhile or even mandatory to implement a complementary approach to the standard concept of a BPM-based energy spectrometer for cross-checking. A promising candidate for such an independent device consists of an upstream spectrometer based on Compton backscattering as outlined in Chapter 6.

Compton spectrometers as employed at low energy storage rings are a priori not practical at the ILC since precise E_b measurements require accurate information based on large event rates. Short bunch crossings of picoseconds duration of high density bunches preclude access to the maximum γ-ray energy out of thousands of backscattered photons. Therefore, the method proposed for the linear collider has to be different and relies, basically, on the determination of the minimum energy of the scattered electrons instead of the maximum energy of the Compton photons.

---

[1]This happens when the B-field scales with E_b which might be not necessarily the case.

3 Magnetic Chicane as Beam Energy Spectrometer

3.1 General Considerations

As discussed in the previous chapter, an energy spectrometer based on a chicane of magnets seems to be the suitable solution for beam energy measurements at a high energy linear collider. A study on the feasibility of such a device was performed some years ago [37]. In this paper a spectrometer composed of three magnets as shown in Fig. 3.1 was proposed.

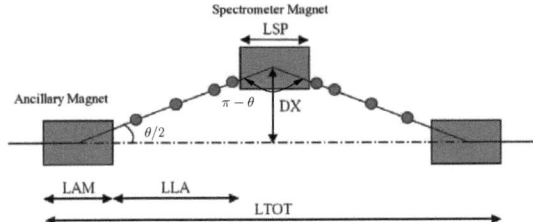

Figure 3.1: Basic layout of a 3-magnet chicane. The circles represent beam position monitors (BPMs).

The aim of the chicane is to measure the deflection angle θ of the beam passing through the spectrometer, see Fig 3.1. To deduce θ, the transverse position of the beam is measured before and after the spectrometer magnet by means of beam position monitors (BPMs). In section 3.4 the working principle of the BPMs is briefly described.

Another option for a magnetic chicane consists of four magnets as shown in Fig. 3.2. In such a scheme, all magnets are identical, but Magnet1 and Magnet4 have an opposite field with respect to that of Magnet2 and Magnet3, allowing the energy to be measured as a function of the induced displacement d in the mid-chicane

$$E_b \propto \frac{L_m}{d} \int Bdl , \qquad (3.1)$$

where $\int Bdl$ is the B-field integral of any magnet and L_m the distance between the first and second magnet.

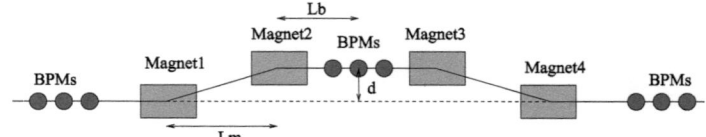

Figure 3.2: Basic layout for a 4-magnet chicane. The circles represent beam position monitors (BPMs).

To determine the displacement d it is necessary to measure two positions of the beam, namely the position of the deflected beam in between Magnet2 and Magnet3 and that of the beam for zero-field, or the position of the extrapolated beam trajectory at the same Z-value (dashed line in Fig. 3.2).

At a circular machine such as LEP, beam position measurements could be performed many times, contrary to the ILC. Hence, at the linear collider the performance of the spectrometer, in particular that of the BPMs, has to be substantially higher. For this reason, resonant cavities were found to be best suitable, because they provide excellent position resolution together with good stability. At the ILC, high resolution BPMs are necessary throughout the main linac and the beam delivery system (BDS). In the linac, the beam position has to be measured with a resolution of about 1 μm for orbit corrections and emittance preservation. In the BDS, dedicated cavity BPMs are required to bring the beams in collision and to ensure high luminosity. Current test facilities and modern linac based sources demand high resolution BPMs. For example, the ATF2 [38] facility at KEK is designed to test the final focus optics of the ILC. Here, cavity BPM position resolutions of 100 nm (or better) with good stability are needed and were partially achieved in the past [39].

A resonant cavity BPM measures the beam offset with respect to the cavity center as well the tilt of the particles traversing the cavity. To minimize the contribution from the tilt, present beam energy spectrometer designs for the ILC propose a 4-magnet chicane of a total length of ∼50 m and an offset of ∼5 mm in the mid-chicane. A BPM resolution of at least 500 nm is demanded in order to achieve a relative beam energy measurement error of 10^{-4} or better. Moreover, a similar BPM stability over hours is mandatory to avoid extensive recalibration and consequently luminosity loss.

A prototype of such a spectrometer was proposed in 2004 and realized during 2006-2007 in End Station A (ESA) at the Stanford Linear Accelerator (SLAC). The basic goal of this device was to study the performance and stability of the key components, namely of the BPMs and bending magnets (experiment T474/T491, see the proposal in Ref. [10]).

This chapter involves the following sections. In Sect. 3.2 an overview of the SLAC linac and End Station A (ESA) is presented. Then, the layout of the experiment T474/T491 with its main components is described and the goals of the experiment are discussed in some details (Sect. 3.3). Section 3.4 summarizes main properties of cavity BPMs, while a detailed description of these devices is beyond the scope of this thesis.

3.2 SLAC Linac and End Station A

3.2.1 SLAC Linac

A basic scheme of the SLAC linac and the beam transport line as used for our experiment is shown in Fig. 3.3. The high intense beam is produced by a thermionic gun, bunched and pre-accelerated to 1.19 GeV in the first section of the linac and then transport to the North Damping Ring (DR), where it is stored for 8 ms to reduce the beam emittance. Afterward, the beam is transported back to the main linac and subsequently accelerated to 28.5 GeV. At the end of the linac, the beam is bent by 24.5° into the A-line on its way to End Station A (ESA), through the Beam Switch Yard (BSY).

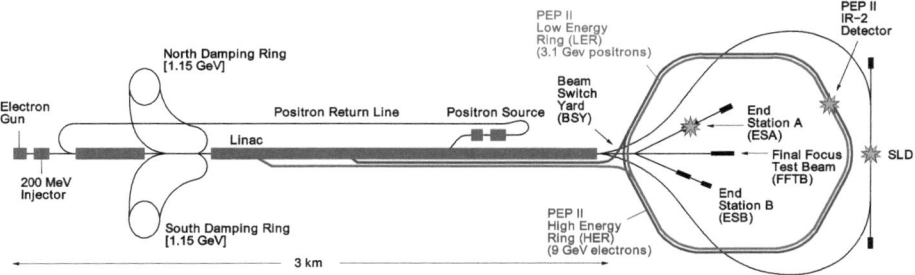

Figure 3.3: Basic layout of the SLAC linear accelerator complex.

3.2.2 End Station A

End Station A is a facility located at the end of the linac, where a single bunch beam with 10 Hz can be delivered to the experiment T474/491 parasitically to PEP-II operation. The main beam parameters of ESA are shown in Table 3.1, where they are also compared with the design parameters for the ILC.

Parameters	ESA	ILC ($\sqrt{s} = 500$)
Repetition rate	10 Hz	5 Hz
Beam energy	28.5 GeV	250 GeV
Train length	up to 400 ns	1 ms
Bunch spacing	$20 \div 400$ ns	337 ns
Bunch per train	1 or 2	2820
Bunch charge	$1.6 \cdot 10^{10}$	$2 \cdot 10^{10}$
Bunch length	500 μm	300 μm
Energy spread	0.15%	0.1%

Table 3.1: Beam parameters at ESA and the values proposed for the ILC (see Sect. 1.1).

In 2006/2007, ESA was the highest energy test facility available with beam parameters similar to those of the ILC. This facility was used to prototype and test major components of the Beam Delivery System (BDS) and the Interaction Region (IR) of the ILC, see [11].

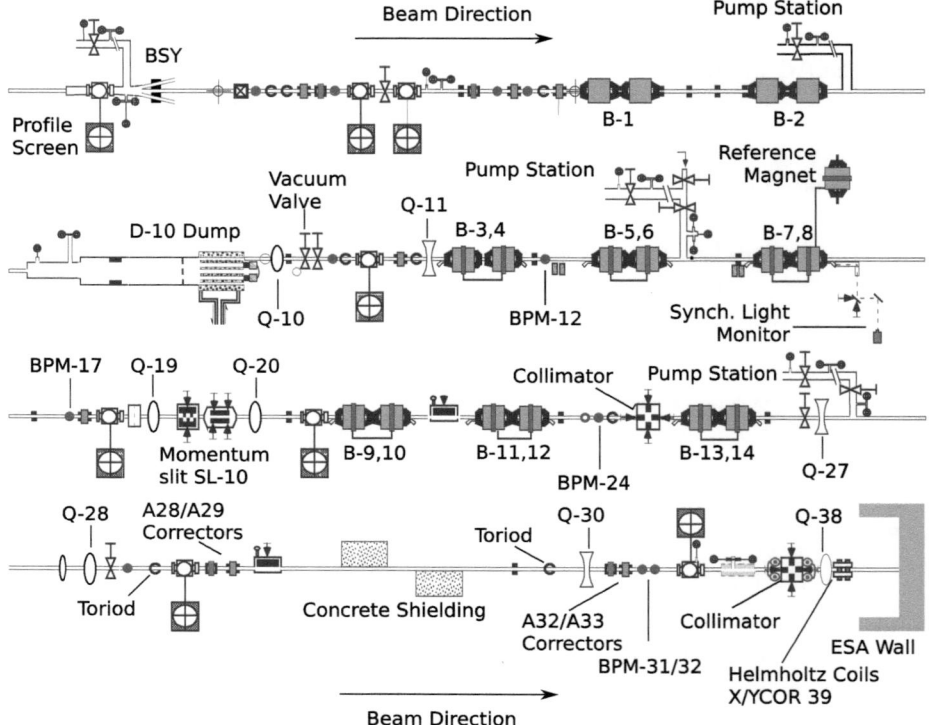

Figure 3.4: Beam line configuration for the A-line up the start of the ESA experimental hall. The beam goes from the left to right, top to bottom

3.2.2.1 The A-Line

The beam from the end of the linac (Beam Switch Yard, BSY) is driven to ESA through the A-line. A schematic representation of the line including its main components is given in Fig. 3.4. After an initial bend of 0.5° in the BSY, a string of 12 dipoles deflects the beam further by 24°.

Three BPMs are used along the bending part in order to monitor the beam position and to provide an energy feedback. In fact, due to the presence of the dipoles, any energy variation is directly reflected into position variation and, as shown later, two of these BPMs (labeled 12 and 24) are used to determine the relative beam energy resolution of the prototype chicane in ESA (see Chapter 5). Other BPMs (31 and 32) are located at the end of the A-line to provide some feedback and corrections, together with two couples of dipoles (A28/A29 and A32/A33).

Upstream of ESA, a couple of Helmholtz coils exist for further beam corrections. Further A-line diagnostics includes charge-sensitive toroids, a synchrotron light monitor, retractable profile screens and high frequency diodes.

3.3 Experiment T474/491

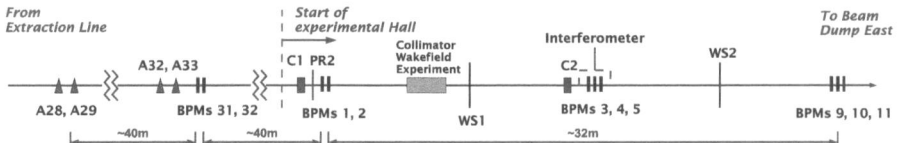

(a) ESA beam line configuration in 2006; no magnets were present.

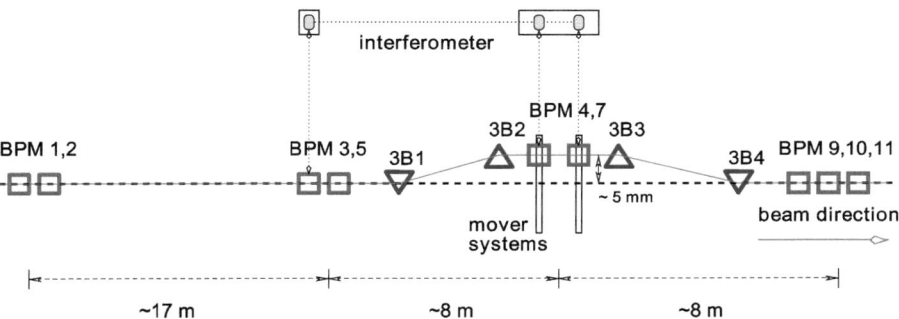

(b) Basic layout of the 4-magnet chicane with the magnets 3B1, 3B2, 3B3 and 3B4 in February 2007.

Figure 3.5

3.2.2.2 ESA Beam Line

The configuration of the ESA beam line together with beam diagnostics and experimental equipment is shown in Fig. 3.5. Two protection collimators are located in ESA: C1 (with 19 mm aperture radius) at the entrance in front of BPM 1 and C2 (with 8 mm aperture radius) located in front of BPM 3. There are two beam profile monitors, one (PR2) upstream of the Collimator Wakefield Experiment and another (PR4, not seen) just beyond the East Wall of ESA. Two wire scanners WS1 and WS2 are placed 20 m apart and used for transverse beam size and emittance measurements. Several beam loss monitors are located along the beam line and interlocked to the machine and radiation protection system.

In ESA the beam is purely ballistic and no optics elements or feedback systems are present.

3.3 Experiment T474/491

During the two-years period 2006/2007 the test experiment T474/491 was installed in ESA in order to study the performance and reliability of a prototype 4-magnet chicane. The prototype was composed of four bending magnets, several BPM stations and an interferometer to control the relative transverse positions of the BPMs.

During 2006, the experiment took data in two runs of two weeks duration each, in April and July, sharing the beam with other experiments (see [12]). The setup involved eight BPMs

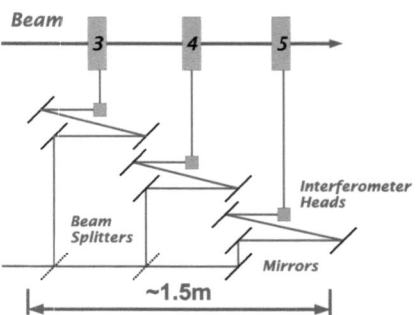 

Figure 3.6: Left: A schematic representation of the laser path in the interferometer relative to BPMs 3-5. Right: The interferometer mounted on the optical table and the BPM triplet 3-5.

mounted on three BPM stations. The first station is composed of the BPMs labeled 1 and 2, the second of the BPMs 3, 4 and 5 and the last one of the BPMs 9, 10 and 11. BPMs installed on a given station are of the same type. The monitors of the first and third station are rectangular cavity BPMs, and each cavity was actually composed of three independent cavities: one for X-position reading, one for Y-position reading, i.e. for transverse position monitoring of the beam, and a reference cavity (see Sect. 3.4). BPMs 1 and 2 were built for the A-line, whereas BPMs 9-11 were initially designed for the main SLAC linac. BPM 3-5 are cylindrical cavities designed for the cryogenic region of the main linac of the ILC. For each of these BPMs, a single cavity provides X- and Y-position measurements. The central monitor BPM 4 is mounted on a mover system with a maximum travel range of 6 mm, due to radiation and machine protection reasons. A three single-pass linear interferometer system provides information of the relative motion along the X-axis, i.e. of the horizontal displacement of the BPM with respect to the interferometer heads as shown in Fig. 3.6.

For the two runs in 2006, no magnets were installed and the main objective of the experiment was to study the performance of BPMs 3-5, in particular their resolution and stability. The results obtained are published in [40].

In an extension of the experiment T474 in 2007 (experiment T491) four bending magnets, labeled 3B1-4, were installed, see Fig. 3.5b. BPM 4 and its mover system were placed in the mid-chicane together with a new BPM, labeled BPM 7, which was also mounted on the mover system. This BPM is a cavity BPM similar to BPMs 3-5, designed and fabricated by the University College London (UCL). As for T474, an interferometer system was able to measure the relative position of BPM 4 and BPM 7 in the mid-chicane with respect to BPM 3 and 5, located upstream of the spectrometer. To study monitor systematics and eventually to infer the absolute beam energy, the magnetic chicane was supposed to run with both B-field polarities. For that, the task of the mover system was essential. As explained in the next section, the linearity of the monitor response drops if the beam passes away ($>1 \div 2$ mm) from the BPM center. Therefore, it is mandatory to position the center of the BPM close to the beam line which is enabled by the mover system.

3.4 Resonant Cavity Beam Position Monitor

A detailed discussion of resonant cavity monitors is beyond the scope of this thesis, but since the BPMs represent an essential part of the energy spectrometer, they will be shortly discussed. Also, since an analysis of the relative resolution of the spectrometer will be presented in Chapter 5, a basic knowledge of these devices is necessary in order to understand the final results. So, in this section an introduction of the basic principles of cavity BPMs is given, together with some general informations on signal processing and calibration procedures.

3.4.1 Resonant Cavity and Beam Coupling

A charged particle traveling within a beam pipe induces a mirror charge in the pipe itself. For a perfect conducting pipe, this charge is traveling together with the beam without loss of energy.

A resonant cavity represents a discontinuity along the beam pipe where stationary waves are induced by the passage of a charged particle. In this case, some energy is stored which oscillates between pure electric and magnetic energy. The total stored energy, W_s, is given as $W_s = <W_e> + <W_m> = 2<W_e>$, where $<W_e>$ and $<W_m>$ are the mean electric and magnetic energy, respectively, averaged over one period. Thus, a cavity can be schematically represented by an LC circuit with a frequency $\omega = 1/\sqrt{(LC)}$ [41]. In other words, a particle passing a discontinuity induces an infinite number of stationary waves n, each of them can be represented by an LC circuit with its own frequency ω_n.

Considering the electric and magnet fields within a cylindrical cavity, we are interested on the fields with pure transverse magnetic oscillations, i.e. on magnetic fields with no longitudinal component ($H_z = 0$). Such fields are denoted as TM modes. They are defined by the geometry of the cavity (length and radius) and by three integer numbers m, n and p. It is common to identify a mode with the notation TM$_{mnp}$ and its frequency is given by

$$\omega_{mnp} = \frac{1}{\sqrt{\mu_0 \epsilon_0}} \sqrt{\left(\frac{j_{mn}}{R}\right)^2 + \left(\frac{p\pi}{l}\right)^2} , \qquad (3.2)$$

where R is the radius of the cavity, l its length and j_{mn} the n-th zero of the m-th Bessel function. For particles near the center of the cavity, the TM$_{010}$ or monopole mode has the

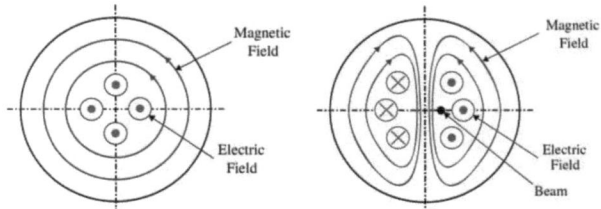

Figure 3.7: Left: Transverse view of the fields of the TM$_{010}$ mode. Right: Transverse view of the fields of the TM$_{110}$ mode. Here, the electric field has a strong dependence on the beam offset from the center of the BPM.

strongest excitation. The explicit expressions for the fields of this mode are [39]

$$E_{z,010} = C_{010} J_0\left(\frac{j_{01} r}{R}\right) e^{i\omega_{010} t} \tag{3.3}$$

$$H_{r,010} = 0 \tag{3.4}$$

$$H_{\phi,010} = -iC_{010}\frac{\omega_{010}\epsilon_0 R}{j_{01}} J_0'\left(\frac{j_{01} r}{R}\right) e^{i\omega_{010} t} . \tag{3.5}$$

The electric field $E_{z,010}$ has a weak symmetric dependence on the distance r from the center. On the other hand, the mode TM$_{110}$ or dipole mode is antisymmetric and its amplitude has a strong dependence on r. The explicit expressions for the dipole fields are [39]

$$E_{z,110} = C_{110} J_1\left(\frac{j_{11} r}{R}\right) \cos\phi \, e^{i\omega_{010} t} \tag{3.6}$$

$$H_{r,110} = -iC_{110}\frac{\omega_{110}\epsilon_0 R^2}{j_{11}^2 r} J_1\left(\frac{j_{11} r}{R}\right) \sin\phi \, e^{i\omega_{010} t} \tag{3.7}$$

$$H_{\phi,110} = -iC_{110}\frac{\omega_{110}\epsilon_0 R}{j_{11}} J_1'\left(\frac{j_{11} r}{R}\right) \cos\phi \, e^{i\omega_{010} t} . \tag{3.8}$$

A schematic representation of both modes TM$_{010}$ and TM$_{110}$ is shown in Fig. 3.7.

As mentioned above, a charged particle traveling through a cavity interacts with the modes and releases some energy. To understand the coupling between the particle and the modes we use the so-called *fundamental theorem of the beam loading*: the voltage induced by a charge traveling through a cavity is twice the effective voltage "seen" by the charge itself [42, 43]. Hence, the energy stored in the cavity by the dipole mode can be calculated as the volume integral of the modulus square of the electric field

$$W_{110} = \frac{1}{2}\epsilon_0 \int_V |E_{z,110}|^2 dV = \frac{\pi}{4}\epsilon_0 C_{110}^2 J_0^2(j_{11}) R^2 l . \tag{3.9}$$

On the other hand, according to the fundamental theorem of the beam loading, the variation

3.4 Resonant Cavity Beam Position Monitor

of the energy stored in the cavity ΔW_{110} can be written as

$$\Delta W_{110} = q \cdot V = \frac{q}{2} \int_{-\infty}^{+\infty} E_{z,110} \vec{v} dt =$$
$$= \frac{q}{2} \int_{-l/2}^{l/2} C_{110} J_1 \left(\frac{j_{11}\delta x}{R}\right) e^{ik_{110}z} dz =$$
$$= \frac{q}{2} C_{110} J_1 \left(\frac{j_{11}\delta x}{R}\right) Tr^{110} l \quad (3.10)$$

for a charge q traveling parallel to the cavity axis with an offset δx, $\phi = 0$ and velocity $\vec{v}$ close to the light speed. Here, the integral is calculated along the path of the particle, q its charge, k_{110} the wave-number and Tr^{110} the transit time factor

$$Tr^{110} = \frac{\left(\int_{-l/2}^{l/2} E_z \cdot e^{ikz} dz\right)}{\left(\int_{-l/2}^{l/2} E_z dz\right)} = \frac{\sin k_{110}l/2}{k_{110}l/2} . \quad (3.11)$$

It is important to note that the phase of the particle with respect to the field induced by the particle itself is chosen such that the field maximally opposes the motion of the particle [43].

Considering the cavity initially empty, Eq. (3.10) is equal to the energy stored (3.9) and approximating the Bessel function $J_1(x)$, for small arguments, by $x/2$, the amplitude C_{110} can be written as

$$C_{110} = \frac{2qTr^{110} J_1 \left(\frac{j_{11}\delta x}{R}\right)}{\pi \epsilon_0 J_0^2(j_{11}) R^2} \approx$$
$$\approx \frac{qTr^{110} j_{11} \delta x}{\pi \epsilon_0 J_0^2(j_{11}) R^3} \quad (3.12)$$

and, in a similar way, C_{010} of the monopole mode

$$C_{010} \approx \frac{qTr^{010}}{\pi \epsilon_0 J_1^2(j_{01}) R^2} . \quad (3.13)$$

Both modes depend linearly on the particle charge q. However, only the dipole mode has a linear dependence on the beam offset. It can be easily shown that two beams with the same δx but opposite in X ($\Delta \phi = \pi$) induce the same voltage with opposite phase. A schematic representation of this behavior is given in Fig. 3.8.

In general, the motion of a particle is not exactly parallel to the Z-axis of the cavity but has an inclination or slope x'. In the linear regime of small offsets, any trajectory can be represented as a sum of a trajectory with only an offset and a trajectory with only an inclination (see Fig. 3.9),

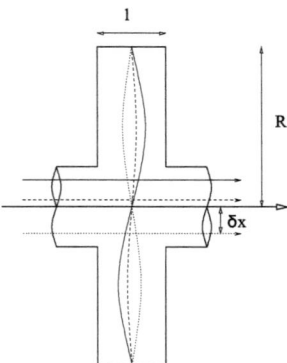

Figure 3.8: Schematic representation of the induced dipole mode in a cylindrical cavity. A beam with small offset (dashed line) induces a smaller signal than a beam with bigger offset (continuous line). Beams with the same offset but opposite in X induce the same signal but with opposite phases (continuous and dotted lines).

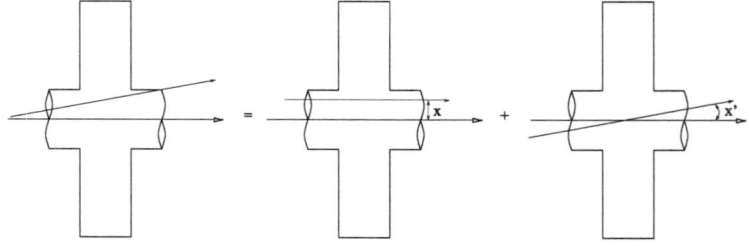

Figure 3.9: A generic trajectory through a cavity interpreted as a superposition of a trajectory with pure offset and a trajectory with pure inclination.

and for the voltage seen by a particle with pure inclination and no offset we have

$$\Delta W_{110,x'} = \frac{q}{2} \int_{-l/2}^{l/2} C_{110} J_1 \left(\frac{j_{11} z \tan x'}{R} \right) e^{ik_{110}z} dz$$
$$\approx i\frac{q}{2} C_{110} \frac{j_{11} x'}{k_{110}^2 R} \left(\sin \frac{k_{110}l}{2} - \frac{k_{110}l}{2} \cos \frac{k_{110}l}{2} \right). \quad (3.14)$$

Thus, the signal is proportional to x'. Comparing this results with Eq. (3.9), the amplitude

$$C_{110} \approx i \frac{2q j_{11} x'}{\pi \epsilon_0 J_0^2(j_{11}) k_{110}^2 R^3 l} \left(\sin \frac{k_{110}l}{2} - \frac{k_{110}l}{2} \cos \frac{k_{110}l}{2} \right) \quad (3.15)$$

reveals that a phase of 90° exists between the field induced by a particle with only an offset and the field induced by a particle with only an inclination. In the next section it will be shown how the two signals can be disentangled.

So far, only cavities with perfect conducting walls were considered. In practice, however, some dissipation of energy in the wall will happen so that the cavity behaves like a RLC circuit

3.4 Resonant Cavity Beam Position Monitor

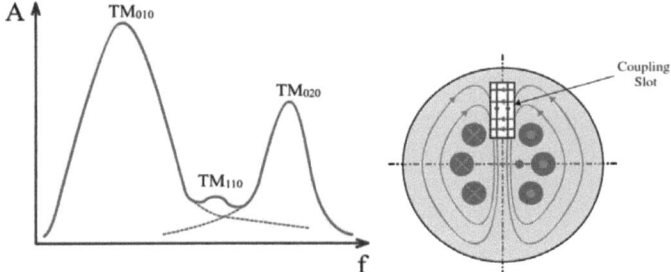

Figure 3.10: Left: Amplitude vs. frequency of the first two monopole modes and the first dipole mode of a cylindrical cavity with non-zero resistivity. The monopoles TM_{010} and TM_{020} surround and overlap the dipole mode TM_{110} [44]. Right: The dipole mode is selectively coupled out by means of a narrow radial slot on one face of the cavity [39].

with a decay constant τ. For this reason, as a Fourier analysis shows, a broad spectrum of frequencies for each mode occurs. In particular, the monopole mode substantially overlaps the dipole mode as indicated in Fig. 3.10(left).

To extract the relevant dipole signal from the cavity, a mode selection is necessary. The mode selection is based on the fact that the boundary conditions for the dipole and monopole modes are different on the wall of the cavity. The dipole mode generates a field transverse to the Z-axis which might have a strong coupling to an opportune modeled waveguide. Thus, it is expected that inside the waveguide the dipole mode is dominant [39, 44] and its amplitude is proportional to the beam offset (see Fig. 3.10(right)).

Finally, a generic dipole mode can be interpreted as a superposition of two orthogonal polarizations of the mode itself. Hence, one cavity can provide X- as well as Y-position readings at the same time (see Fig. 3.11).

Figure 3.12 (left) shows a photograph of a cylindrical cavity BPM. It is BPM 7 installed in the mid-chicane of the experiment T474/491. The cavity, as can be seen, is connected with four waveguides, two in vertical direction to extract the X-position signal and two in horizontal direction for the Y-position signal.

Similar conclusions can be drawn for rectangular cavities. An example of such a cavity, BPM 9 is shown in Fig. 3.12 (right). The main difference to the cylindrical cavity is that X- and Y-position reading is performed by two distinct cavities, the two rectangular cavities in the picture.

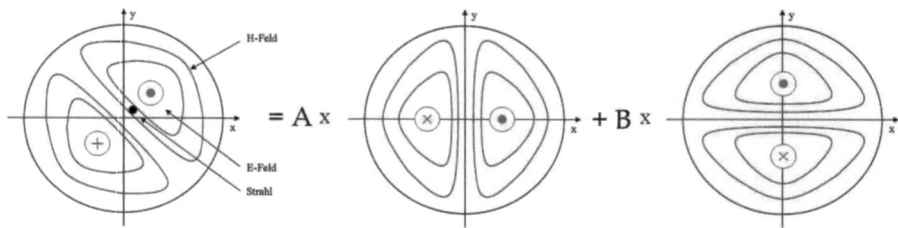

Figure 3.11: A generic dipole mode can be represented as a superposition of two orthogonal polarizations of the mode.

Figure 3.12: Left: Cylindrical cavity BPM as designed by UCL. Right: BPM 9 in End Station A. The two rectangular cavities for X- and Y-position reading and the cylindrical reference cavity are clearly seen.

3.4.2 Signal Processing

It is common to express the output voltage of the dipole mode as a function of the shunt impedance and quality factor. The shunt impedance is defined as

$$R_{110} = \frac{V_{110}}{P_{110,\,loss}} \quad , \tag{3.16}$$

where $P_{110,loss}$ is the power dissipated in the cavity walls. The internal quality factor is defined as

$$Q_{110}^{int} = \frac{w_{110} W_{110}}{P_{110,\,loss}} \quad . \tag{3.17}$$

Thus the energy stored in the cavity can be written as

$$W_{110} = \frac{\omega_{110}}{4} \left(\frac{R}{Q}\right)_{110} q^2 \quad , \tag{3.18}$$

3.4 Resonant Cavity Beam Position Monitor

where $\left(\frac{R}{Q}\right)_{110}$ is the normalized shunt impedance. This quantity is independent on the material of the cavity and depends only on its geometry. Moreover, it has a finite value also when the cavity has a wall with zero resistivity (which corresponds to an infinite value of the internal quality factor).

According to Eqs. (3.9) and (3.12), we have $\left(\frac{R}{Q}\right)_{110} \propto (\delta x)^2$. Defining the external quality factor as

$$Q_{110}^{ext} = \frac{w_{110}W_{110}}{P_{110,\,out}}, \tag{3.19}$$

the dipole output power is then given by

$$P_{110,\,out} = \frac{\omega_{110}}{4Q_{110}^{ext}} \left(\frac{R}{Q}\right)_{110} q^2 . \tag{3.20}$$

Finally, the readout electronics with impedance Z provides an output voltage of

$$V_{out} = \sqrt{ZP_{110,\,out}} = \frac{q\omega_{110}}{2} \sqrt{\frac{Z}{Q_{110}^{ext}} \left(\frac{R}{Q}\right)_{110}} . \tag{3.21}$$

As can be seen, this voltage has a linear dependence on the charge and the offset of the particle, denoted as x in the following. In order to measure a signal which is only proportional to the offset, normalization to the charge is needed. For this reason the monopole signal is simultaneously extracted from a reference cavity, which is tuned such that its monopole mode has the same frequency as the dipole mode of the beam position cavity, i.e. $\omega_{010}^{ref} = \omega_{110}^{BPM}$. Moreover, the reference cavity provides the arrival time of the beam, allowing to determine the phase of the signal.

The voltage signal at the front-end of the analogue electronics has thus the form

$$\begin{aligned} V(t) &= e^{-\Gamma t}[A_x x \sin \omega t + A_{x'} x' \cos \omega t] \\ &= a e^{-\Gamma t} \sin(\omega t + \phi) , \end{aligned} \tag{3.22}$$

where $a = \sqrt{(A_x x)^2 + (A_{x'} x')^2}$ and $\phi = \arctan(A_{x'} x'/A_x x)$, with amplitudes A_x and $A_{x'} \propto q$. In a similar way the reference cavity provides

$$V(t) = a_{ref} e^{-\Gamma t} \sin(\omega t + \phi_{ref}) , \tag{3.23}$$

where $a_{ref} \propto q$.

After filtering and digitization, both signals are multiplied by a complex local oscillator (LO) of the same frequency as the signal. This process, called digital down-conversion (DDC), results

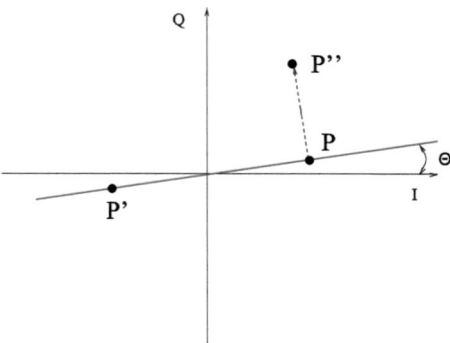

Figure 3.13: I-Q plot with an IQ-phase Θ. Changing the position of the beam, the point P moves along the continuous line and changes sign when it crosses the cavity center. Changing the tilt, P moves along the dashed line, perpendicular to the continuous line.

in a signal which describes the envelope of the initial waveform

$$V_{DDC}(t) = ae^{-\Gamma t}\sin(\omega t + \phi) \cdot e^{i\omega t}$$
$$= \frac{a}{2}e^{-\Gamma t}\left[-e^{i(\phi + \frac{\pi}{2})} + e^{i(2\omega t + \phi + \frac{\pi}{2})}\right]. \quad (3.24)$$

The mixing process also generates an unwanted wave with a frequency 2ω, which has to be eliminated by an additional filter. Customarily, the quantities I and Q denote the real, respectively, imaginary part of the normalized down-converted signal

$$I = \frac{a}{a_{ref}}\cos(\phi - \phi_{ref}) \quad (3.25)$$

$$Q = \frac{a}{a_{ref}}\sin(\phi - \phi_{ref}). \quad (3.26)$$

An example of an I-Q plot is shown in Fig. 3.13 for a beam particle with a generic offset and tilt generated at the point P. If the tilt (or slope) is kept constant while the position is changed, P moves along the continuous line, which has, in general, a non-zero slope Θ (IQ-phase). When the beam crosses the center of the BPM, P changes sign (point P'). If on the other hand, the beam position is kept unaltered while the beam tilt is changed, the point P moves along the dashed line to point P", perpendicular to the continuous line.

To extract position and tilt of the particle, the IQ-phase Θ and the scale factors S and S' have to be determined by an appropriate calibration procedure. Finally, the position and tilt

3.4 Resonant Cavity Beam Position Monitor

are obtained after rotation of the plane by angle Θ and multiplication with the scale factors:

$$x = S \cdot Re\left[\frac{a}{a_{ref}}e^{i(\phi-\phi_{ref}-\Theta)}\right] \qquad (3.27)$$

$$x' = S' \cdot Im\left[\frac{a}{a_{ref}}e^{i(\phi-\phi_{ref}-\Theta)}\right] . \qquad (3.28)$$

In the experiment T474/T491, calibration was performed by generating a well-known beam offset using the corrector magnets or the Helmholtz coils of the A-line (see Sect. 3.2.2.1 and [40]). For BPMs 4 and 7, also the mover systems have been utilized for calibration purposes. In ESA, only the offset and not the tilt could be calibrated.

4 Characterization of the Magnets

4.1 General Considerations

The 4-magnet chicane in ESA is composed of two major parts, the beam position monitors to measure the transverse position of the beam and the dipole magnets. Knowledge of the B-field integral ($\int Bdl$) is indispensable quantity to determine the energy of the beam. Therefore, it is important to determine and monitor the fields with best precision as possible during the runs. The relative accuracy needed for $\int Bdl$ was estimated to be approximately 50 ppm.

For the T-474/491 experiment, old SPEAR magnets labeled as 10D37 were slightly refurbished and put into the beam line. But before installation, detailed measurements were performed at the SLAC laboratory between December 2006 and February 2007 to study stability and reproducibility of the magnets as well as to monitor the $\int Bdl$ over long time [45, 46].

In this chapter field measurement techniques as well as instruments for B-field and B-field integral determinations together with the results obtained are presented. The chapter is organized as follows. Section 4.2 summarizes main parameters of the SPEAR magnets. In Sect. 4.3 an overview of the measurement techniques is given including a description of the setup in the laboratory and the types of measurements performed. Section 4.4 presents the results of the measurements, while Sect. 4.5 discusses the error on B-field integral determinations in details. After a summary of the measurements (Sect. 4.6), some suggestions are given in Sect. 4.7 how to improve the results obtained. Finally, Sect. 4.8 describes the implementation of the magnets in ESA.

4.2 The Magnets 10D37

The magnets were built in 1971 to deliver the beam to the SPEAR ring. They have a length of 94 cm and a weight of about 1100 kg, see Fig. 4.1. A maximum field of about 1 T can be reached. The magnets are solid steel H-type warm magnets with a small but not well defined fraction of carbon, estimated to be less than 0.13%. They were properly stored at SLAC for more than 10 years, so that only routine cleaning but no special refurbishing was needed for our experiment. The magnets are symmetric with respect to the longitudinal and transverse planes but two slightly different mirror plates were added to contain fringe fields. The magnets are labeled 3B1, 3B2, 3B3 and 3B4 according to their positions within the beam line.

Figure 4.1: Left: Magnets 10D37 in the laboratory with mirror plates. Right: A sketch of magnet 10D37 without mirror plates.

4.3 B-field Measurements and Techniques

The quantities measured in the laboratory were the vertical component of the B-field inside and outside of the gap of the magnets, the B-field integral ($\int Bdl$) and the magnet current.

The magnetic chicane was planned to operate with both polarities in order to perform absolute energy measurements and to study possible spectrometer systematics. Therefore, it is important to understand the stability of the field, the $\int Bdl$ and the magnet current over long time scales (of hours or days), especially after switching the polarity. Also, correct standardization is important since otherwise, for example, $\int Bdl$ monitoring can be disturbed, as discussed later.

The mapping of the B-field along the beam direction using one or more probes mounted on a moving arm was also worthwhile for comparison of the data with presimulations [47] and, in turn, to use the measurements as an input for improved simulations.

However, the most important point is to monitor $\int Bdl$ for which a procedure was developed to deliver this quantity during data taking runs in ESA. We decided to perform $\int Bdl$ monitoring using NMR probes, since no other adequate device was available to measure directly the B-field integral.

An NMR probe measures the field in a particular point and not $\int Bdl$. We rely on the basic assumption that when the B-field changes by some amount in a fixed point within the magnet, every field point changes by the same amount. In other words, we assume that the point related B-field is proportional to $\int Bdl$. The choice of this point will be discussed later (see Sect. 4.3.2) and by means of a procedure described in Sect. 4.4.6 the coefficient of this proportionality is evaluated so that finally the field integral, $\int Bdl$, is accessible.

4.3.1 Instruments

In general, to perform B-field and $\int Bdl$ measurements the following devices are commonly utilized:

- NMR probe, for absolute B-field measurements.

4.3 B-field Measurements and Techniques

- Hall probe, for relative B-field measurements.
- Flux gate, for absolute low B-field measurements.
- Moving wire, for absolute $\int Bdl$ measurements.
- Flip coil, for relative $\int Bdl$ measurements.

4.3.1.1 NMR Probe

NMR probes are usually composed of a sample of water wrapped around with a coil. When an external magnetic field is applied, the spins of the protons are aligned to the direction of the field and precess with a frequency of $\nu_L = \gamma |B|/(2\pi)$, known as the Larmor frequency. γ is the gyromagnetic ratio and B the external field. Under such conditions, the proton can have two energy levels, one with spin orientation parallel and one anti-parallel to the field with energy separation proportional to the Larmor frequency. If a weak, perpendicular oscillating B-field is applied by a coil with a frequency equal to the Larmor frequency, transitions from spin-parallel states (lowest energy) to those with spin anti-parallel (highest energy) can be observed. The value of ν_L can be measured very precisely from which the B-field can be calculated. Advantages of this device are high precision, absolute and non-axial field measurements, while disadvantages concern the need of uniform fields (the gradient must be smaller than $6 \cdot 10^{-7}$ T/m), a limited working range, non-applicability for wrong field orientation and no information on the sign of the field [48, 49].

4.3.1.2 Hall Probe

Considering a conductor with a current passing through it and a B-field applied perpendicular to the current vector $\vec{J}$, according to the Lorentz force an accumulation of negative charge (which corresponds to an opposite accumulation of positive charge) is observed in the plane where $\vec{J}$ lies. The equilibrium is reached when the electrostatic force between the accumulated negative and positive charges is equal to the Lorentz force. For this case, the electrostatic force is proportional to the external B-field and is measured. Advantages of this device are a large working range, insensitivity to non-uniform or time dependent B-fields and information on the field direction. Disadvantages are the need of calibration, reduced precision and only the field component perpendicular to $\vec{J}$ is measured.

4.3.1.3 Flux Gate

The flux gate is a device suitable to measure low fields. Several flux gate configurations exist, but all of them are based on the following principle (see Fig. 4.2): two ferromagnetic bars are wrapped by a common coil in opposite direction where an AC current passes through. In cases of no external field, the field induced in one bar compensates exactly the other one in the second bar, resulting in a total null-flux in a second coil which wrapped both bars. If, however, an external constant field parallel to the bars is present, a time dependent flux is generated which induces a voltage in the second coil. Advantages of the flux gate are high precision for low fields and absolute field measurements. Possible disadvantages are the dependence on the field orientation and its non-applicability to fields above 4 Gauss.

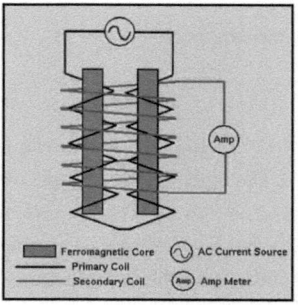

Figure 4.2: A simple scheme of a flux gate device.

4.3.1.4 Moving Wire Technique

A ribbon of wires is passed through the gap of the magnet and closed outside to form a circuit. The wires are secured at the end of each side of the magnet at holders mounted on a mover stage. Moving the wires along the transverse direction X, a voltage is induced and integrating this voltage the B-field integral $\int Bdl$ can be deduced from the Eq. (4.1)

$$\int Bdl = \frac{\int Vdt}{N\Delta x}, \qquad (4.1)$$

where V is the voltage, N the number of turns of the wire and Δx the distance traveled by the mover stage. The advantage of this device consists in an absolute $\int Bdl$ measurement and disadvantages are the slow rate, possible instability of the mover stage over long time periods and misalignment errors [33].

4.3.1.5 Flip Coil Technique

The scheme of the flip coil technique is shown in Fig. 4.3. A coil is inserted in the gap of the magnet perpendicular to the dipole field. If the coil is rotated over 180 degrees, a voltage is induced and if this voltage is integrated in time (like for the moving wire), the B-field integral $\int Bdl$ can be inferred from

$$\int Bdl = \frac{\int Vdt}{2dN}, \qquad (4.2)$$

where V is the voltage induced, N number of the wire turns and d the effective diameter of the coil. Advantages of this instrument are high rate measurements resulting in a smaller

4.3 B-field Measurements and Techniques

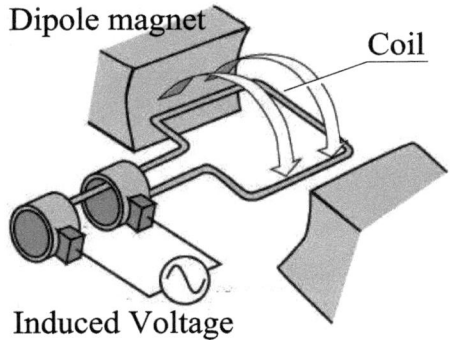

Figure 4.3: A simple scheme of the flip coil technique.

statistical error and high stability. Disadvantages may be the need of calibration and errors due to misalignment [33].

4.3.2 Experimental Setup

The measurements of the 10D37 magnets were performed in the magnet laboratory at SLAC using an aluminum test bench table (Fig 4.4). For all four magnets, different measurement procedures as described below were applied, which result in different types of data discussed at the end of this section.

The system of reference for the measurements was defined as follows. The point X=Y=Z=0 corresponds to the geometrical center of the magnet. The Z-axis points along the longitudinal axis of the magnet, the X-axis to the horizontal direction and the Y-axis is chosen such that a right-handed system is obtained, see also [47] and Fig. 4.1 (right).

Before each measurement, the magnets were standardized to stabilize the field by three ramping cycles. The current was ramped up and down between -200 and +200 A, which corresponds to the maximum possible values provided by the power supply in ESA.

In a first step, an additional magnet with high field homogeneity was used to calibrate the Hall to the NMR probe, whereas the flip coil was calibrated to the moving wire device using one of the 10D37 magnets.

To measure $\int B dl$ with the moving wire technique, the wire was moved in X-direction by 1 cm, covering the center of the magnet. The 1 cm driving distance corresponds roughly to the rod diameter of the flip coil.

Five measurements with the moving wire have been done and averaged, while for the flip coil the average was obtained from 32 measurements, rotating the coil 8 times clockwise and 8 times counter-clockwise. For the analysis presented in Sect. 4.4, the average values measured from the moving wire and flip coil techniques were taken into account and denoted as "single measurement". A measurement with the flip coil or moving wire took about $1 \div 2$ min, while probe measurements were much faster. The NMR probe, for example, can sample every $1 \div 2$

sec. For comparison of flip coil with probe data, the probe values were also averaged over the corresponding time period of a flip coil/moving wire measurement.

Figure 4.4: One of the 10D37 magnets on the test bench table during a B-field map. The beam pipe and one of the mirror plates mounted on one side of the magnet are also visible.

For stability and reproducibility data taking periods, a short beam pipe and mirror plates were mounted on the magnets in order to achieve conditions as in ESA. The NMR probe was fixed close to the pipe through an appropriate holder at a position where the field gradient, obtained from simulation, was smaller than $6 \cdot 10^{-7}$ T/m. The Hall probe was placed directly above the pole face and the flux gate was fixed on the beam pipe outside the magnet. For low field measurements, such as the residual field inside the magnet, only the Hall probe and the flux gate were used and both were placed inside the magnet.

For field mapping, the NMR and Hall probes and the flux gate were mounted at the end of a mover arm system. Using a stepper motor, it was possible to move them precisely along X-, Y- and Z-directions.

Before performing Hall probe measurements, the probe was zeroed by inserting it into a μ-metal pipe, where a very low field of 0.065 Gauss was present.

Stability and reproducibility runs as well as field maps were performed at 200 and 150 A, corresponding to field values of 0.15 and 0.11 T, respectively.

In summary, the following types of measurements were performed:

- Stability runs: the magnet 3B4 was powered with a fixed current over a long period, typically over 6÷24 hours. We measured $\int Bdl$ using the flip coil technique and, independently, NMR and Hall probes for the B-field of a particular point inside the magnet.

- Reproducibility runs: the magnets were operated by flipping the polarity periodically so that the reproducibility could be studied by monitoring $\int Bdl$. This is very similar to the way to operate the magnets in ESA. These measurements were performed for magnets 3B1, 3B2 and 3B4.

- Current scan: B-field and integrated B-field measurements were performed by ramping the current in fixed steps over a large or small current range. It was thus possible to study the dependence of the B-field integral versus the current or NMR probe measurements, for example. These runs were performed for magnets 3B1, 3B2 and 3B3, while for magnet 3B4 only a large range current scan was carried out.

- Field maps along X- and Z-directions at zero-current and working currents were performed. The Z-mapping at the working current was done for all magnets, while X-mapping and mapping at zero-current were only performed for magnets 3B1 and 3B4.

4.4 Results of B-field Measurements

4.4.1 Field Mapping

4.4.1.1 Field Mapping in X-Direction (X-Scan)

Figure 4.5 shows the results of the field mapping in X-direction at Y=Z=0, i.e. in the center of the magnets 3B1 (left) and 3B4 (right) using Hall and NMR probes, while in Fig. 4.6 the simulation results [47] are displayed for X> 0 only, since it was assumed that the B-field is symmetric with respect to X=0. Comparing the two figures within the appropriate X-range, we conclude that simulation and measurements agree rather well.

As can be noticed from Fig. 4.5, the minimum value of the field, the magnetic center, is not positioned at the geometric center of the magnet but shifted by -7 mm. The measured values of the shift for magnets 3B1 and 3B4 are summarized in Tab. 4.1.

Magnet	Hall Probe (mm)	NMR Probe (mm)
3B1	-7 ± 2	-8 ± 2
3B4	-6 ± 2	-7 ± 2

Table 4.1: Measured shifts of the magnetic center along X at Y=Z=0.

4.4.1.2 Field Mapping in Z-Direction (Z-Scan)

The behavior of a magnetic field in Z-direction is usually represented as a step function, which is, strictly speaking, not true since outside the magnet the B-field approaches smoothly to zero. This fraction of the field is called fringe field.

In a first step, we are interested to understand the impact of the mirror plates on the shape of the most important fringe field component B_y. Some basic arguments suggest that the fringe field can introduce some non-linearity between the B-field in a particular point within the magnet and the B-field integral (see Sect. 4.4.1.3).

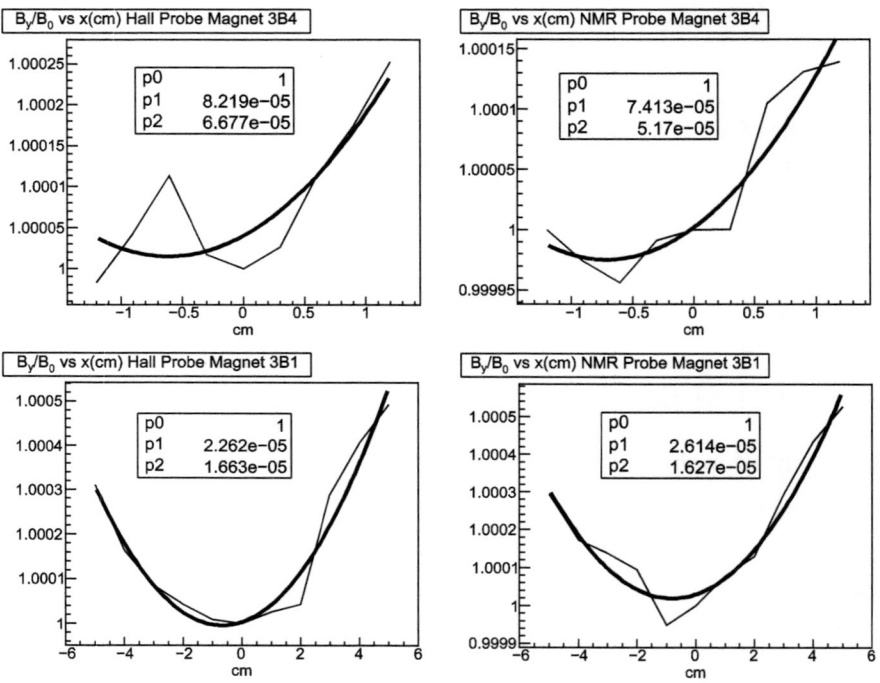

Figure 4.5: B-field measurements from X-scans with NMR and Hall probes at Z=0 for magnets 3B1 and 3B4. The continuous curves are the results of a fit with a parabola.

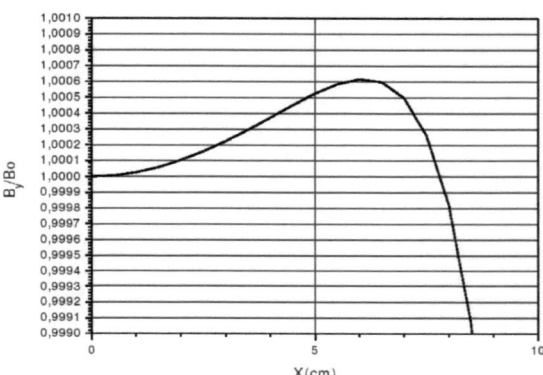

Figure 4.6: B-field simulation versus X at Z=0. B_0 is the value of the B-field for Z=X=0.

4.4 Results of B-field Measurements

For magnet 3B1, several Z-scans near both ends of the magnet were performed, with and without mirror plates and with and without the beam pipe. Figure 4.7 shows the results on one side of magnet 3B1, with and without the mirror plates. It is obvious that the mirror plates have a strong impact to reduce substantially the fringe field.

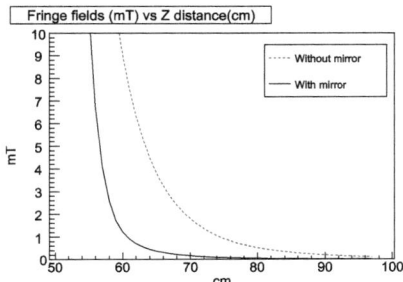

Figure 4.7: Z-scans of the fringe field of magnet 3B1 with (continuous) and without (dotted) mirror plates. The abscissa represents the Z-coordinate.

A comparison of the Z-scan with simulations is displayed in Figs. 4.8 and 4.9 for magnet 3B1. The Z-scan was performed at both ends of the magnet (Run3 and Run5), with the beam pipe and the mirror plates installed. Only Fig. 4.9, after zooming the fringe field region, clearly reveals differences between simulations and measurements. The maximum difference amounts to approximately 0.7 mT, which was considered by the collaboration as acceptable taking into account the assumption made in the simulation. Furthermore, due to a slight asymmetry of the mirror plates, small differences are present between the fringe fields on each side of the magnet.

4.4.1.3 Z-Scan for Zero-Current Fields

Using the flux gate and Hall probe, a Z-scan at zero-current was performed in order to measure the residual magnetization of the magnet and the stray field. The latter one is supposed to be a superposition of the earth field and any B-field produced by surrounding cables and electronic equipment.

As shown in Figs. 4.10 and 4.11, the residual field for magnets 3B3 and 3B4 is about 0.4 mT. One notices from Fig. 4.10 that the data at Z-values near 45 and 60 cm are outside the working range of the flux gate device of some 4 Gauss. The right-hand side of Figs. 4.10 and 4.11 shows a zoom of the tail of the field for Z-values larger than 90 cm. Here it is supposed that the residual B-field from magnet is negligible and the stray field is dominant. The points shown in Fig. 4.10 are measurements performed with both the flux gate and the Hall probe, while only Hall probe measurements are shown in Fig. 4.11. As can be seen, the stray field amounts to $0.012 \div 0.017$ mT, in agreement with independent measurements presented in [50].

We also note that in Fig. 4.11 the sign of the field measured is opposite to that in Fig. 4.10. This is due to an error which appeared after data taking: the flux gate and Hall probe were installed with wrong orientations and no attention was paid to correct for this. Such a situation

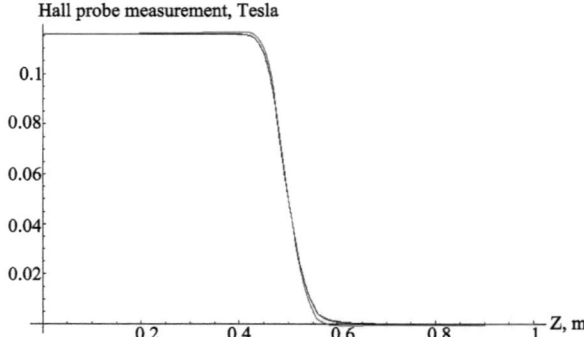

Figure 4.8: Comparison of Z-scan data with simulations for both end sides of magnet 3B1. In the figure, Hall probe measurements of the vertical B-field component are plotted as function of the distance Z from the magnet center.

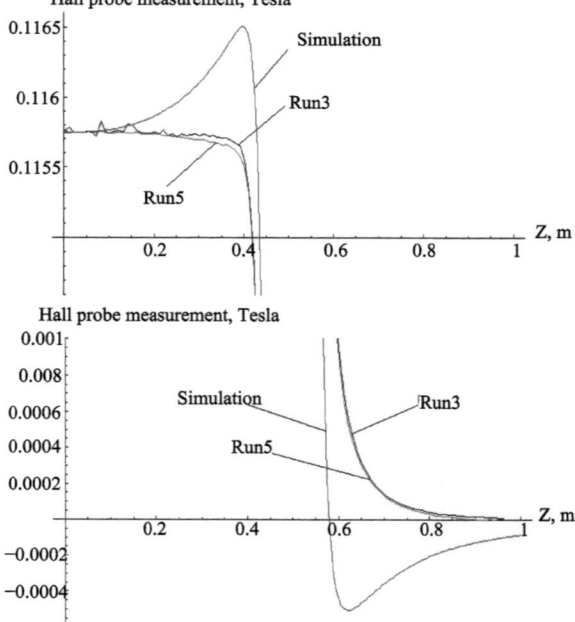

Figure 4.9: A zoom of Fig. 4.8 in the fringe field region reveals some differences between the scan data and simulations. Run3 corresponds to the upstream end scan, while Run5 to the downstream scan.

4.4 Results of B-field Measurements

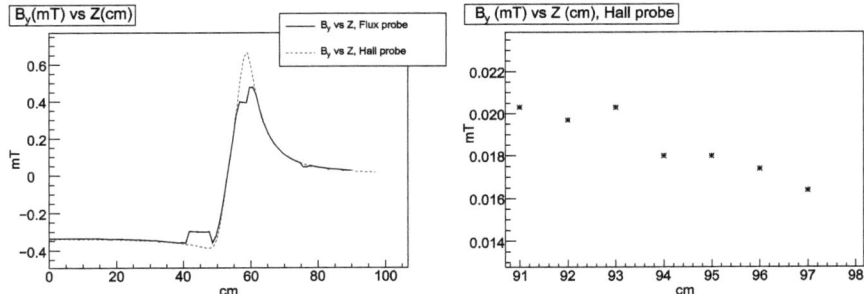

Figure 4.10: Left: Z-scan at zero-current for magnet 3B4. Right: A zoom of the tail where the stray field is dominant.

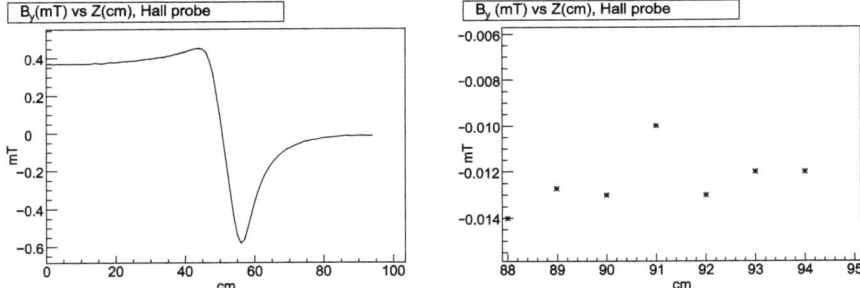

Figure 4.11: left: Z-scan at zero-current for magnet 3B3. Right: A zoom of the tail where the stray field is dominant.

happened also for other measurements. For example, in a reproducibility run of magnet 3B4 the signs of the flip coil, Hall probe and flux gate measurements coincide, while for magnet 3B1 the flip coil measurement has an opposite sign compared to Hall probe and flux gate data. For this reason only absolute values of B-field and B-field integral measurements will be considered for the data analysis, since it is too difficult to correct for it.

It is interesting to see that the shape of the fields in Figs. 4.11 and 4.10 is different to the fields in Figs. 4.8 and 4.9. This supports the supposition that the fringe field can introduce a non-linearity between the B-field in a particular fixed point within the magnet and the B-field integral.

4.4.2 Field Stability Runs

For magnet 3B4 some stability runs were performed. After the standardization procedure, the magnet was powered with a fixed current of +150 A or +200 A over long periods ($6 \div 24$ hours) and data were recorded from the flip coil ($\int Bdl$), NMR and Hall probes (B-field).

Figures 4.12a and 4.12b show the relative variation of NMR and Hall probe data as well as data from the flip coil and magnet current as functions of the sample number for a 24 and 6 hour run, respectively. We define the magnet stability as the RMS of the $\int Bdl$ measurements. For the 24 hour run, the stability was found to be close to 30 ppm, whereas for the 6 hour run some step behavior of the probe and coil measurements can be seen. This peculiarity is not observed for the current measurements. Since no further data exist, we can only suppose that this behavior might (probably) be due to magnet temperature variations (for details see Sect. 4.4.3).

4.4.2.1 Residual of Stability Runs for Magnet 3B4

For the 24 hour run, we can evaluate the ratio

$$P1 = \frac{<\int Bdl>}{<B>} = 0.9965 \pm 0.0003 \ , \tag{4.3}$$

where $<\int Bdl>$ is the average of flip coil B-field integral measurements and $<B>$ denotes the average of NMR probe measurements.

$P1 \cdot B_{NMR} = (\int Bdl)_{pred}$ provides a prediction for the B-field integral. The difference between this prediction and the corresponding flip coil measurement, $(\int Bdl)_{pred} - (\int Bdl)_{flip}$, reveals how well the NMR probe represents $\int Bdl$. We call this difference as residual. Figure 4.13 shows the distributions of the residuals for the 24 hour run (left) and 6 hour run (right), respectively. For the 24 hour run, the RMS is about 16 ppm (about the half of the stability run value), whereas for the 6 hour run it is close to 41 ppm. This increases from 16 to 41 ppm can only be explained qualitatively: if the working conditions, such as the temperature of the magnet, changes quickly for unknown reasons, the linear relation between the fixed point B-field and B-field integral is violated and the RMS of the residuals can increase.

4.4.2.2 Stability Run for Magnet 3B4 at Zero-Current

For a special zero-current run the residual field inside magnet 3B4 was monitored using the flux gate, the Hall probe (installed inside the gap of the magnet) and the flip coil. In Fig. 4.14 the relative variations of each measurement are presented against the sample number, together with the cooling water temperature. The data shown were taken over 6 hours. As can be seen, a variation of the residual B-field in the order of 0.1% exists, which is accompanied by a strong dependence on the cooling water temperature. The Hall probe shows a systematic fall-off (with large fluctuations) which might be ascribed to an insufficient precision of the probe for very low fields.

The absolute residual field of the magnet 3B4 (not shown) is about 0.3 mT.

4.4.3 Temperature Dependence

Figure 4.15 shows $\int Bdl$ and the B-field for magnet 3B4 measured by the flip coil and NMR probe versus the temperature of the magnet pole.

4.4 Results of B-field Measurements 53

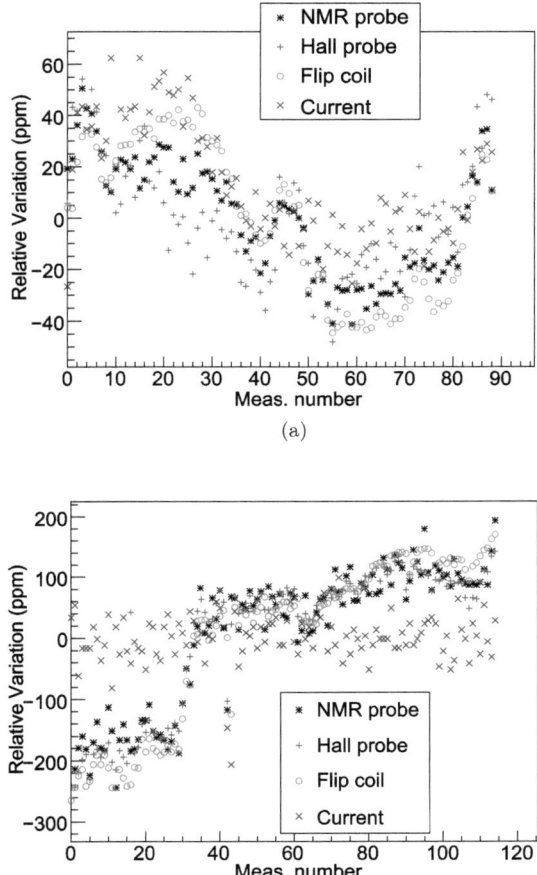

Figure 4.12: Relative variations of the flip coil, Hall and NMR probe measurements and magnet current values for two stability runs of magnet 3B4: a) for 24 hours; b) for 6 hours.

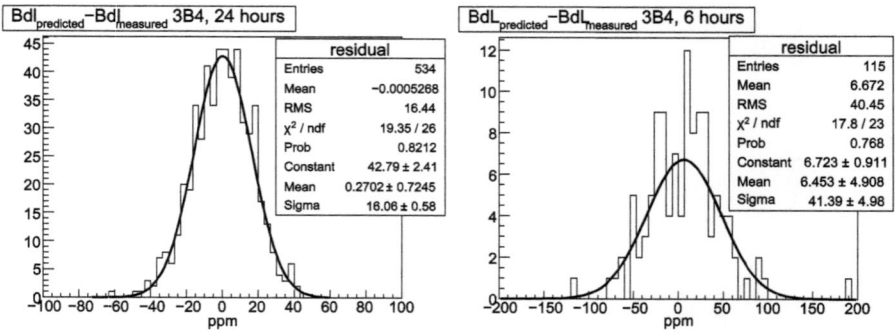

Figure 4.13: Residuals of the 24 hour (left) and 6 hour run (right) for magnet 3B4.

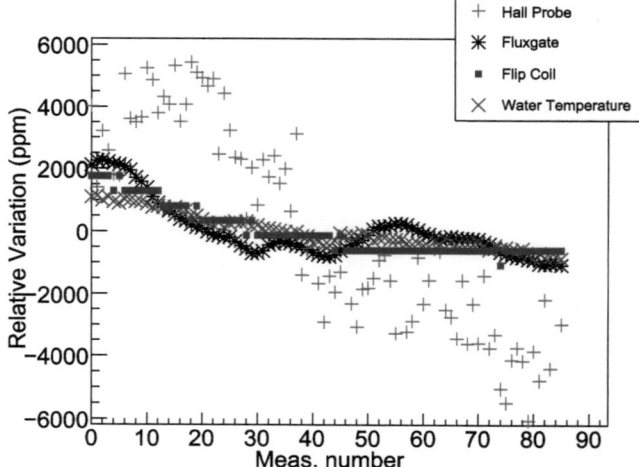

Figure 4.14: Results of the residual field measurements of magnet 3B4.

According to simulations [47], a linear dependence between the $\int Bdl$ and the temperature is expected. In [47], the $\int Bdl$ variation on temperature is due to a change of the magnetization curve and due to changes of the dimensions of the magnet. The simulation estimated temperature factor was found to be $6.1 \cdot 10^{-5}$ °C^{-1}.

Performing a simple linear fit to the data in Fig. 4.15, a temperature factor of

$$F_{\int Bdl} = (4.2 \pm 1.0) \cdot 10^{-5} \text{ °C}^{-1} \tag{4.4}$$

is found which we consider to be in good agreement with the simulation value from [47].

It is also useful to know the correlation factor between the B-field and the temperature, which was estimated to be

$$F_{B-field} = (2.2 \pm 0.5) \cdot 10^{-5} \text{ °C}^{-1} \ . \tag{4.5}$$

4.4 Results of B-field Measurements

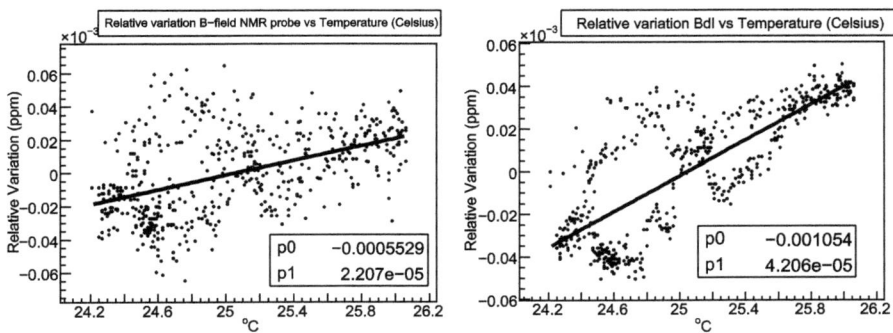

Figure 4.15: Dependence of the relative B-field (left) and B-field integral (right) on the temperature. Some measurements have different values at the same temperature because of current fluctuation.

For each factor a conservative error of 20% was assigned. The NMR probe has, compared to the flip coil, a different temperature factor which has to be considered when the NMR measurements are used to calculate the B-field integral.

4.4.4 Reproducibility Runs

At ESA, it was planned to run the magnets with both polarities in order to measure the beam energy absolutely in situ. For this reason, the behavior of the magnets by flipping the polarity was studied in some details.

Figure 4.16 shows the results of $\int Bdl$ for magnet 3B1 from reproducibility runs. The current was ramped up and down between +150 and -150 A and for each polarity, data were taken with the NMR and Hall probes, placed inside the magnets, as well as the flux gate, placed outside of the magnet, and the flip coil. The field polarity was changed periodically 11 times per run, and for each time, 6 measurements of $\int Bdl$ and the B-field were performed, resulting in 6 samples of 6 measurements for a given polarity (see left-hand side of Fig. 4.16).

An adequate quantity to characterize the magnet behavior is the difference between the fields of negative and positive polarity.

We calculated the mean value of each sample in the left-hand side of Fig. 4.16. Afterward, the difference between two consecutive mean values with opposite polarity was determined and on the right-hand side of Fig. 4.16, the differences normalized to the modulus of the B-field are plotted. An RMS value of about 40 ppm is observed.

Moreover, the absolute mean values of the B-field for negative and positive field orientation were not identical (not visible in Fig. 4.16). An 'offset' of $0.2 \div 0.4$ mT was observed, which might be due to some residual field, being measured to be non-zero after switching off the current.

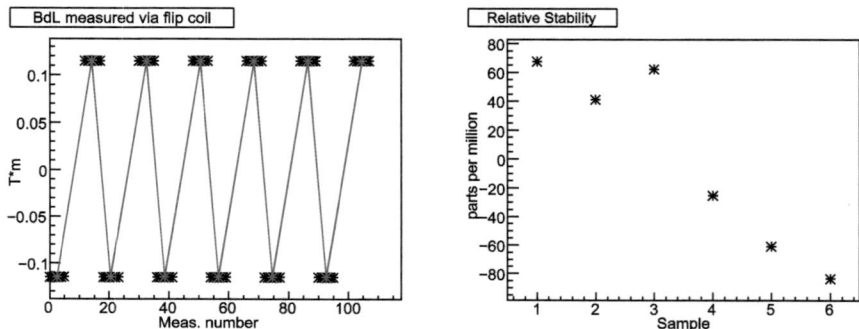

Figure 4.16: Results of reproducibility run for magnet 3B1.

4.4.5 Summary from Stability and Reproducibility Runs

Here we summarize some conclusions from the stability and reproducibility runs. For the 24 hour data taking period, a stability of 30 ppm and an RMS of the residuals between NMR and flip coil data of 16 ppm were found. A slightly worse result was obtained from the 6 hour run, with an RMS of the residuals of about 42 ppm. As already stated, no explanation is given for the jump in Fig 4.12b and we do not know how to control it. Both RMS values are however acceptable. Also, the reproducibility run was found to be reasonably stable, with a relative variation of the difference between the B-fields of negative and positive polarities in the order of 40 ppm. These results confirm that the standardization procedure was able to stabilize the magnets reasonably well and the NMR probe under such conditions predicts the B-field integral with an acceptable error which does not exceed the requirement of 50 ppm (see Sect. 4.1).

4.4.6 Current Scans of Magnets

Current scans were performed for all magnets over large and small ranges. For large range scans, the current was changed from -200 to +200 A and then back to -200 A in 25 A increments, while small range scans were performed between +140 and +150 A in 1 A steps. Figure 4.17 displays the results for the scan between -200 and +200 A of magnet 3B4.

The slope of a straight line fit results in

$$a1 = 0.000781 \pm 0.000016 \text{ T} \cdot \text{m} \cdot \text{A}^{-1} \tag{4.6}$$

and was used to determine approximately the B-field integral from the magnet current. From the right-hand side of Fig. 4.17 we notice a maximum $\int Bdl$ error of about 0.2%, if the current is used to predict the field integral.

In order to deduce the B-field integral from the NMR probe we need the coefficients P1 and

4.4 Results of B-field Measurements

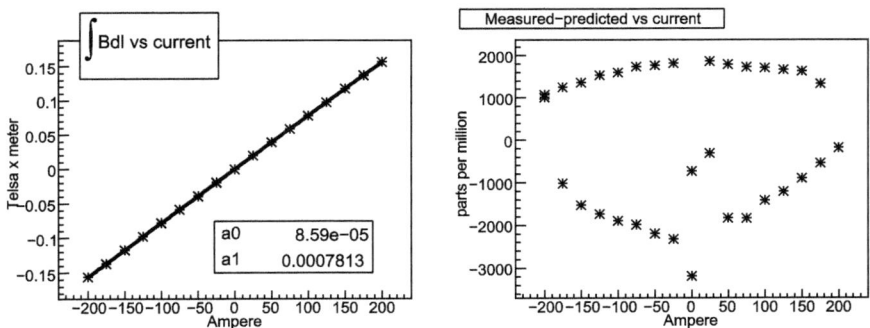

Figure 4.17: Results of the current scan for magnet 3B4. Left: $\int Bdl$ values together with a linear fit as a function of the current. Right: Difference between the $\int Bdl$ fit value and the corresponding measured value. The upper points are the measurements increasing the current from -200 up to + 200 A, while the lower points are obtained when decreasing the current from +200 down to -200 A.

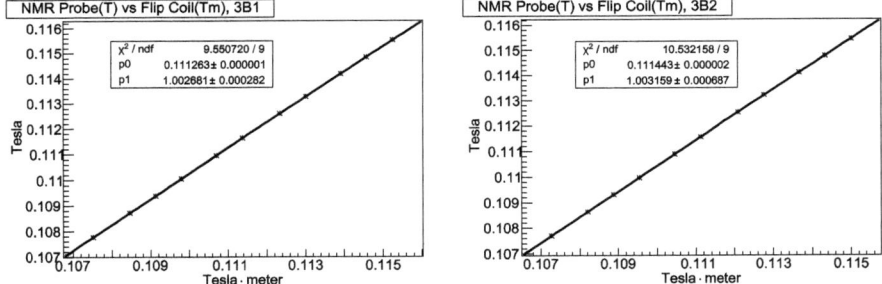

Figure 4.18: Calibration of the NMR probe for magnet 3B1 (left) and magnet 3B2 (right).

Magnet	Slope (m^{-1})	Intercept (T)
3B1	1.0027 ± 0.0003	0.111263 ± 0.000001
3B2	1.0032 ± 0.0007	0.111443 ± 0.000002
3B3	1.0050 ± 0.0008	0.111380 ± 0.000002

Table 4.2: Calibration coefficients of the NMR probe for magnets 3B1 to 3B3

P0 of the linear function

$$\int Bdl = \text{P1} \cdot (B_{NMR} - 0.11) + \text{P0} , \tag{4.7}$$

which is supposed to represent a good approximation. For that, a positive current scan within 140 ÷ 150 A with a 1 A increment was performed. For each current, the B-field from the NMR probe and the B-field integral from the flip coil were recorded and displayed in Figs. 4.18 and 4.19. Measurement fluctuations are due, in this case, mainly to current fluctuations. Since flip coil data are obtained from averaging over a larger number of measurements than the NMR

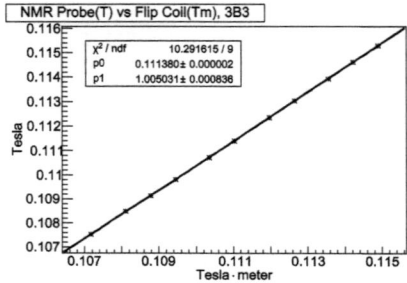

Figure 4.19: Calibration of the NMR probe for magnet 3B3.

probe, the relative fluctuations of the flip coil data were smaller in size than the NMR probe data. For this reason, NMR data were plotted as function of the flip coil data and the inverse of Eq. (4.7) was used for the fit. Table 4.2 collects the slopes and the intercepts for magnets 3B1, 3B2 and 3B3. For magnet 3B4, a corresponding current scan was not performed and the coefficients needed were obtained from the 24 hour stability run, i.e.

$$P1 = \frac{<\int Bdl>}{<B>} = 0.9965 \pm 0.0003$$

and $P0 = 0$ (see Sect. 4.4.2.1). In the following, the procedure to determine the coefficients P0 and P1 is referred to as NMR calibration.

4.4.7 Residuals for Magnets 3B1, 3B2 and 3B4

After calibration of the NMR probe as described in the previous section, reproducibility runs for magnets 3B1, 3B2 and 3B4 were performed. The important quantity here is the residual which is defined as the difference $(\int Bdl)_{pred} - (\int Bdl)_{flip}$, where $(\int Bdl)_{pred}$ is the predicted B-field integral from the NMR probe, $(\int Bdl)_{pred} = P1 \cdot B_{NMR} + P0$, and $(\int Bdl)_{flip}$ the flip coil measurement (see Sect. 4.4.2.1 and 4.4.6).

The reproducibility runs provide six groups of six measurement for each polarity, see Fig. 4.16 and Sect. 4.4.4. For each group, the mean and the RMS of the residuals were calculated and the results are shown on the right-hand side of Figs. 4.20, 4.21 and 4.22. The histograms on the left-hand side display the residual distributions for negative and positive currents of magnets 3B1, 3B2 and 3B4. First, we notice that the mean residuals for negative current are definitely not centered at zero. This can only partially be explained by the stray field as schematically shown in Fig. 4.23. A magnet with a constant vertical field component B_y inside the magnet provides outside a rapidly decreasing field, the fringe field. But outside the magnet an additional field exists which is supposed to be a superposition of the earth field and the B-fields produced by surrounding cables and/or electronic equipment (stray field). The field integral measured by the flip coil or the moving wire is then the sum of two fields

4.4 Results of B-field Measurements

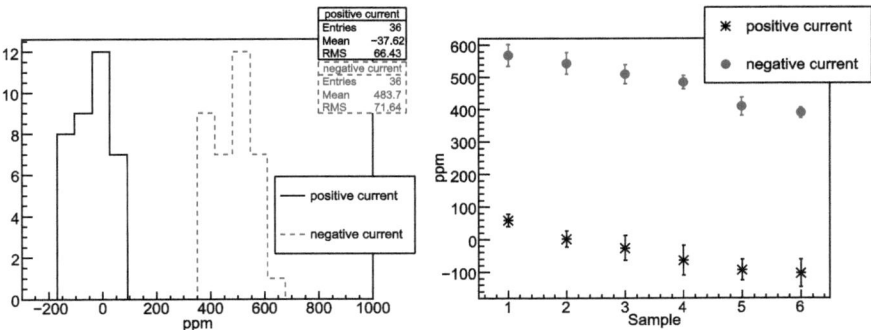

Figure 4.20: Left: Residuals for magnet 3B1 from the so-called reproducibility run, normalized to the mean value of $\int Bdl$. Right: Mean values of the residuals together with the RMS for magnet 3B1 calculated for each sample at both polarities.

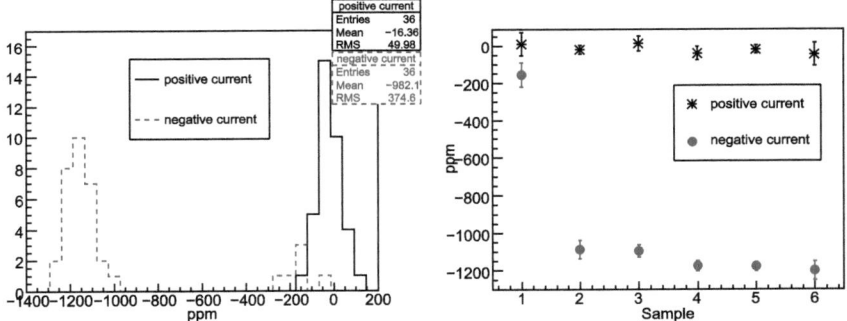

Figure 4.21: Left: Residuals for magnet 3B2 from the so-called reproducibility run, normalized to the mean value of $\int Bdl$. Right: Mean values of the residuals together with the RMS for magnet 3B2 calculated for each sample at both polarities.

$$\begin{cases} |\int Bdl|_{measured} = |\int Bdl|_{magnet} - |\int Bdl|_{stray}, \text{ for positive current} \\ |\int Bdl|_{measured} = |\int Bdl|_{magnet} + |\int Bdl|_{stray}, \text{ for negative current} \end{cases} \quad (4.8)$$

Equations (4.8) follow from the fact that when the polarity of the magnet is flipped, $(\int Bdl)_{measured}$ changes sign but not $(\int Bdl)_{stray}$. Therefore, in one case they sum up while in the other they are subtracted from each other. The minus sign assigned to the positive current was chosen arbitrarily. The strength of $(\int Bdl)_{stray}$ is estimated to be in the order of $(1 \div 2) \cdot 10^{-5}$ T·m. As noted in Sect. 4.4.1.3, it was not possible to determine the sign of the fields, since several errors were made during data taking.

As explained in Sect. 4.4.6, the NMR probe was calibrated from flip coil measurements using

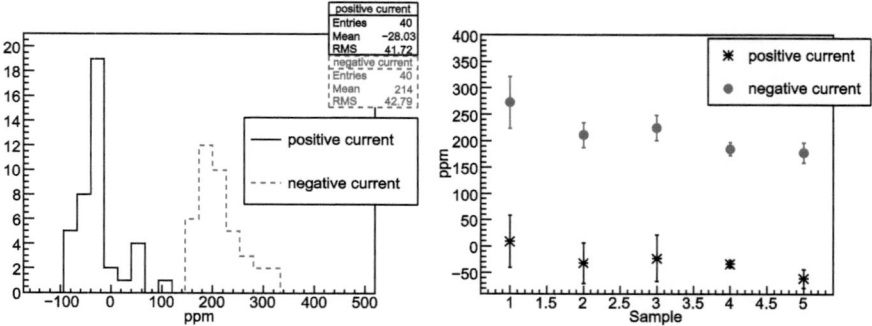

Figure 4.22: Left: Residuals for magnet 3B4 from the so-called reproducibility run, normalized to the mean value of $\int Bdl$. Right: Mean values of the residuals together with the RMS for magnet 3B4 calculated for each sample at both polarities.

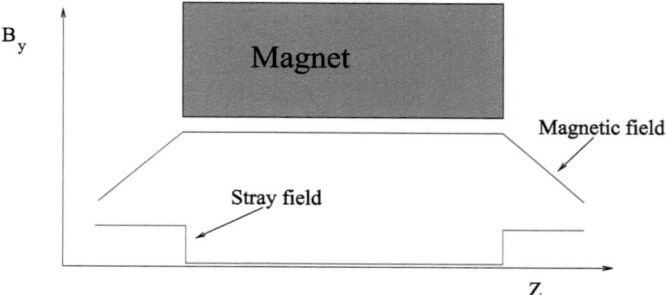

Figure 4.23: Schematic representation of the field component B_y of a magnet and the stray field along the Z-axis

data of positive current runs. From Eqs. (4.8) and (4.7) follows

$$\left|\int Bdl\right|_{measured} = \left|\int Bdl\right|_{magnet} - \left|\int Bdl\right|_{stray} = P1|B_{NMR}| + P0 \quad (4.9)$$

and the residual is calculated as the difference

$$P1|B_{NMR}| + P0 - \left|\int Bdl\right|_{measured} . \quad (4.10)$$

For positive current runs, the mean residual results in

$$(P1|B_{NMR}| + P0) - \left|\int Bdl\right|_{magnet} + \left|\int Bdl\right|_{stray} = 0 , \quad (4.11)$$

whereas, taking into account the expression above and Eq. (4.8), the corresponding mean

4.4 Results of B-field Measurements

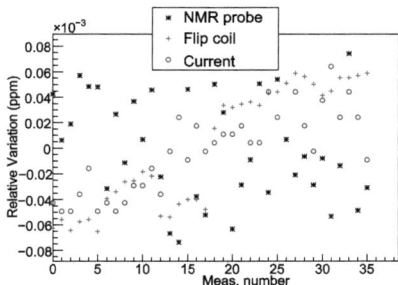

Figure 4.24: Relative variation of current, NMR probe and flip coil values of magnet 3B1 for positive polarity during the reproducibility run.

residual for negative current runs is expressed as

$$(P1|B_{NMR}| + P0) - \left|\int Bdl\right|_{magnet} - \left|\int Bdl\right|_{stray} =$$
$$(P1|B_{NMR}| + P0) - \left|\int Bdl\right|_{magnet} + \left|\int Bdl\right|_{stray} +$$
$$-2 \cdot \left|\int Bdl\right|_{stray} = -2 \cdot \left|\int Bdl\right|_{stray} . \quad (4.12)$$

The results of (4.11) and (4.12) are qualitatively in accord with the data shown in Figs. 4.20, 4.21 and 4.22. If we exchange the signs in Eqs. (4.8), the fields sum up for positive current data and are subtracted for negative ones. This implies a reversed sign in (4.12), which, however, does not modify our final results. The contribution of the stray field to the total B-field integral was estimated to be $(\int Bdl)_{stray} \approx 1.35 \cdot 10^{-5}$ T·m, which corresponds to a fraction of $100 \div 200$ ppm of the total field integral, in agreement with measurements from Figs. 4.20, 4.21 and 4.22.

As best seen from the right-hand side of Fig. 4.20, a drift in time of the mean residual exists. To understand this, Fig. 4.24 shows as an example the relative variations of NMR, $\int Bdl$ and current data of the reproducibility run for magnet 3B1 at positive polarity. The current and $\int Bdl$ flip coil measurements drift with time, but the field measured by the NMR probe does not despite strong fluctuations. This means that the NMR probe is blind with respect to $\int Bdl$ and current variations resulting in the drift seen in Fig. 4.20 (right). If we compare the residuals in Fig 4.20 (left) with those obtained for magnet 3B4 from the stability run (Sect. 4.4.2.1), we note that the latter residuals have a smaller RMS (Fig. 4.13).

The last point of discussion concerns the "jump" in Fig. 4.21 for negative current data when going from sample 1 to sample 2. Since this step cannot be explained by neither current nor temperature variations, the following explanation might be appropriate. Figure 4.25 displays the variations of the Hall probe, NMR probe and flip coil measurement normalized to their mean value of the reproducibility run of Fig. 4.21 for positive (left) and negative currents (right), respectively. For the left-hand side (positive current), a gentle or small jump can be noticed for all three devices, resulting in an acceptable residual. For the negative current,

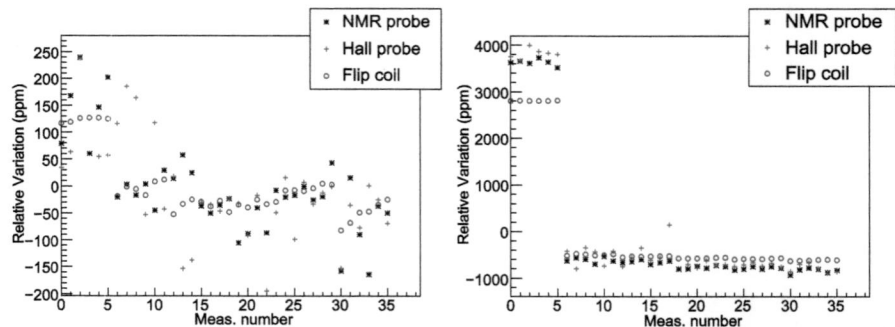

Figure 4.25: Relative variations of NMR and Hall probe measurements as well as $\int Bdl$ data of the flip coil from the reproducibility run of magnet 3B2 for positive (left) and negative (right) current.

the large jump of the flip coil corresponds to a similar behavior of the Hall and NMR probe data but smaller in size. An ideal situation would be when the relative variations of the 3 devices were the same in size (compare e.g. with Figs. 4.12a and 4.12b). Since the two probes provide independent measurements (different devices in different locations) the most probable explanation of the behavior in Fig. 4.21 is a malfunctioning of the flip coil.

4.5 Error Sources and Estimations

In this section we discuss possible error sources associated with each device utilized during the course of the measurements.

An important error might be caused by possible rotations around any coordinate axis (roll, pitch, yaw) and by displacements of the measurement device with respect to the chosen reference system. Unlike the NMR probe, all devices perform axial measurements, i.e. they only measure the field component perpendicular to some plane. So, if we denote $B = B_y$ as a field of only the Y-component, rolling or pitching modifies this field to $B' = B\cos\theta$, and the field error from the uncertainty of θ is $\delta B' = B\delta(\cos\theta) = B\sin\theta \cdot \delta\theta$. $\delta\theta$ is estimated to be less than 0.5 mrad which yields for the worst situation $\theta = \delta\theta = 0.5$ mrad, a relative field error of $\delta B'/B = \sin\theta \cdot \delta\theta = 2.5 \cdot 10^{-7}$, i.e. of 0.25 ppm.

For any displacement it is worth to note that the most sensible axis is the Y-axis. A displacement in Y of ± 1 cm away from the center of the magnet results in a decrease of the field by about $6 \cdot 10^{-4}$, whereas a corresponding displacement along X increases the field by only $(0.5 \div 1) \cdot 10^{-4}$ from its nominal value. For displacements along the Z-axis, B-field and B-field integral variations are negligible as long as measurements are performed close to the center of the magnet.

As already pointed out, B_y depends in general on X-, Y- and Z-positions within the magnet.

4.5 Error Sources and Estimations

Considering a magnet with an effective length l, B_y can be parametrized as

$$B_y(X,Y,Z) = \begin{cases} B_0(p_0 + p_1 x + p_2 x^2)(1 - A|y|) & \text{if } |z| < l/2 \\ 0 & \text{otherwise} \end{cases} \quad (4.13)$$

Here, B_0 is the value of B_y at the magnet center X=Y=Z=0. The parameters p_0, p_1 and p_2 are determined from X-scans of magnet 3B1 (Fig. 4.5) and the parameter A was measured to be $6 \cdot 10^{-4}$ from the field mapping. So, the actual field measured by the flip coil and moving wire techniques is

$$\frac{\int B_0(p_0 + p_1 x + p_2 x^2)(1 - A|y|) \cdot u_y dx dz}{\Delta x} = \frac{\int V dt}{\Delta x} = <B \cdot l>, \quad (4.14)$$

with u_y the vertical component of the normal to the plane defined by the moving wire or flip coil. From the small value of $\sin\theta \cdot \delta\theta$, see above, we can set $u_y = 1$. In absence of yaw, pitch and roll rotations as well as displacements in X- or Y-direction, the field measured by the moving wire or flip coil is reduced to

$$\frac{(\int B_0(p_0 + p_1 x + p_2 x^2) dx) \cdot l}{\Delta x} = \frac{\int V dt}{\Delta x} = <B \cdot l>, \quad (4.15)$$

and the relative difference between the nominal B-field integral ($B_0 l$) and the measurement $<B \cdot l>$ results in

$$\frac{<B \cdot l> - B_0 l}{B_0 l} = 1.5 \cdot 10^{-6}. \quad (4.16)$$

A displacement in X of 2 mm or a yaw rotation of 0.5 mrad modifies this value in a negligible manner. For pitch rotation of 0.5 mrad, the difference becomes

$$\frac{<B \cdot l> - B_0 l}{B_0 l} = 3.4 \cdot 10^{-6}, \quad (4.17)$$

while a 0.5 mrad roll rotation yields

$$\frac{<B \cdot l> - B_0 l}{B_0 l} = 1.2 \cdot 10^{-6}. \quad (4.18)$$

The geometrical sum of the errors due to rotations, non-uniformity of the B-field and displacements in X becomes $\delta(\int Bdl)/\int Bdl = 3.9 \cdot 10^{-6}$, which holds for the flip coil as well as the moving wire device.

The last point to discuss concerns the residuals. The RMS of this important quantity measures how well a device, for example the NMR probe, agrees with values from other methods, e.g. the flip coil technique in our case. The resulting mean value of the residuals is then interpreted as the error of the NMR calibration procedure (see Fig. 4.20).

It is important to remark that calibration and misalignment errors of the flip coil and misalignment errors of the moving wire are not visible in Figs. 4.20, 4.21 and 4.22 and they were calculated above. Misalignment errors of the NMR probe are considered to be negligible. In this way, the final error for the $\int Bdl$ monitoring using the NMR probe will be the quadratic sum of the misalignment error of the moving wire and flip coil, the accuracy of each device, the calibration error of the flip coil and the RMS and the mean value of the residual $(\int Bdl)_{pred} - (\int Bdl)_{flip}$.

The main error sources of the techniques applied can be summarized as:

1. NMR probe:
    - the precision of a single measurement is 5 ppm,
    - the RMS of the residuals [NMR - (flip coil)] is 66 ppm,
    - the error of the NMR calibration is 38 ppm.

2. Hall probe:
    - the precision of a single measurement is 100 ppm + 0.05 Gauss (taken from the manual),
    - the pitching error of the mover arm during Z- or X-scan is less than 0.5 mrad yielding an error of 0.25 ppm,
    - the errors on rolling or pitching position is <0.5 mrad, which corresponds to an error of less than 0.25 ppm.

3. moving wire:
    - the moving wire error of a single measurement is 26 ppm,
    - a displacement error in Y-direction of 2 mm corresponds to a field error of 120 ppm,
    - the error due to B-field non-uniformity, a displacement error in X of 2 mm and a 0.5 mrad uncertainty in any rotation results in 3.9 ppm.

4. flip coil:
    - the resolution of a single measurement is 5 ppm,
    - a displacement error of about 2 mm in Y-direction yields a field error of 120 ppm,
    - the error due to B-field non-uniformity, a displacement error in X of 2 mm and a 0.5 mrad uncertainty in any rotation results in 3.9 ppm.
    - the RMS residual [(moving wire) - (flip coil)] and the [(moving wire) - (flip coil)] calibration error are unknown.

5. current measurement
    - the accuracy for a single measurement is 100 ppm and in addition,
    - a systematic error of -0.01 A has to be assumed.

4.5.1 Error of B-field Integral Monitoring

We conclude this section by evaluating the error on $\int Bdl$ monitoring. The evaluation is performed for magnet 3B1, with data taken for positive polarity during a reproducibility run. Since all the errors discussed above are independent, the quadratic sum of the moving wire, flip coil and NMR probe errors provide the total uncertainty on B-field integral monitoring of

$$\frac{\delta\left(\int Bdl\right)}{\int Bdl} = 184 \text{ ppm.} \qquad (4.19)$$

We note that the largest contribution to this error comes from the alignment error along the Y-axis.

4.6 Summary

Simulations of the 10D37 magnet for the ESA energy spectrometer and the laboratory measurements agree quite well. Largest differences were observed in the region of the fringe field of about 7 Gauss. Moreover, we found that the center of the field does not coincide with the geometrical center of the magnet, a result not reproduced by simulations. The reason of that is unknown.

The total system (magnet+power supply) was found to be very stable: long-term runs showed a stability of $\int Bdl$ and of the current at the level of or better than 30 ppm. This also means that the applied standardization procedure works reasonably well.

Concerning the measurement of the temperature coefficient, the results are not fully convincing despite good agreement with simulations. Fluctuations due to current ripple were of the same amount as temperature fluctuations, spoiling a clear temperature dependence of the B-field. In future, temperature variations must be monitored in order to measure $\int Bdl$ with a relative accuracy of 50 ppm.

Regarding monitoring of the integrated B-field, the NMR probe was able to provide good estimations for $\int Bdl$, except for one peculiarity seen in Fig. 4.21, which is not fully understood (see also Fig. 4.25).

For the stability run (Sect. 4.4.2.1), the (NMR - flip coil) residuals delivered some better RMS than the reproducibility run (compare Fig. 4.13 with Figs. 4.20, 4.21, 4.22).

Flux gate measurements performed outside the magnet and data from the Hall probe, placed in the gap of the magnet, could not be used to evaluate $\int Bdl$. In the first case, the sensitivity of the measurements to variations of the field integral was very weak since the flux gate only measures low fields, e.g. the fringe field, whereas the errors of the Hall probe were too big.

The final error on the B-field integral of 184 ppm is still quite large compared to the requirement of 50 ppm (see Sect. 4.1), but some improvements are possible, especially for the alignment of the devices along the X- and Y-axes. Some additional measurements and optimization procedures to reduce the B-field error are suggested in the next section.

4.7 Recommendations for Future Measurements

In future, additional measurements should be performed for magnets 3B1 and 3B2, since they are the most important magnets in the spectrometer. We suggest to perform detailed measurements using the moving wire and flip coil techniques, by placing the magnets onto an appropriate aluminum test bench table as done for the first round of measurements. In particular, we require a precision alignment of 0.5 mrad along the rotation axes (yaw, pitch, roll) and of 0.1 mm in X and Y for the flip coil and moving wire systems. Also a careful standardization procedure should be adapted, which has to consist of three cycles between -200 and 200 A and be applied before any local field or $\int Bdl$ measurements.

Measurements of the temperature coefficient should be performed after standardization runs with a magnet current of +150 A. Measuring $\int Bdl$ with the flip coil and the field in a fixed point with the NMR probe, the temperature should be varied in the range of $10 \div 20$ °C by adiabatically reducing the water flow. From such measurements it should be possible to deduce the temperature coefficient with an error of 10%.

To reduce the RMS of the residuals between NMR probe and flip coil measurements, we suggest to install more than one NMR probe into each magnet. Their calibration with a current scan between $140 \div 150$ A with 1 A increment for both polarities allows to reduce significantly the errors of the probes.

Since a measurement with the flip coil takes about $1 \div 2$ minutes, several NMR measurements of $1 \div 2$ seconds can be performed within this time period. We propose to record as many NMR measurements as possible during flip coil data taking. So far, a flip coil measurement was supplemented by $3 \div 4$ NMR probe measurements.

Also the residuals between the flip coil and moving wire techniques should be measured. In principle, measurements with the flip coil should be simultaneously performed with the moving wire system by positioning the wire over the flip coil or next to it. However, due to some non-uniformity of the B-field along X or Y, the moving wire will measure a slightly different $\int Bdl$ than the flip coil. Positioning the moving wire next to the flip coil seems to be preferable since the field gradient is smaller along the X-axis. The possible difference between the B-field integral from the moving wire and the flip coil can be calculated using, for example, the data of Fig. 4.5.

Another option to reduce the total error on $\int Bdl$ consists of omitting the flip coil and to calibrate the NMR probe directly with the moving wire device.

Since the time needed for a reproducibility or stability run was typically larger than 6 hours, more care is needed to avoid or reduce alignment errors of the moving wire system.

4.8 The 4-Magnet Chicane in End Station A

Figure 4.26 shows a schematic view of the 4-magnet chicane in ESA. The magnets were installed on a girder, using supports made of non-ferromagnetic material. Only one power supply was

4.8 The 4-Magnet Chicane in End Station A

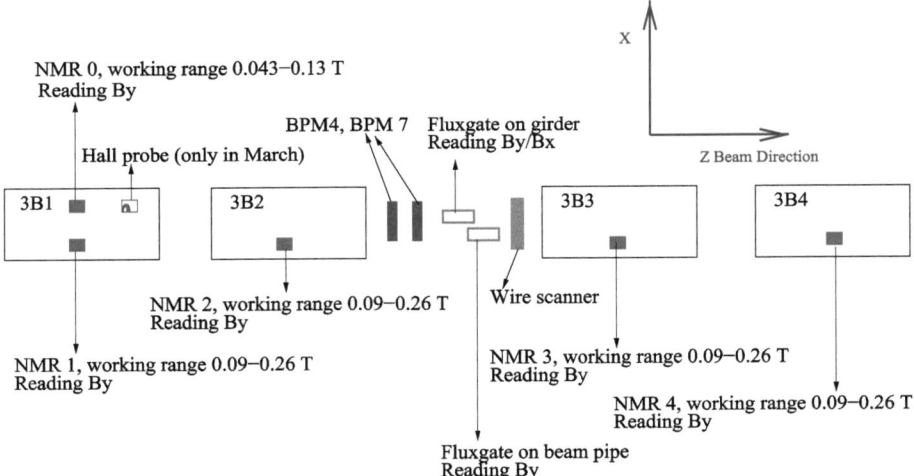

Figure 4.26: Schematic representation of the magnetic chicane in ESA. Also main diagnostics instruments are shown.

available for the magnets mounted in series, so that the current was the same for all of them. By means of a Linux computer it was possible to set the nominal values of the B-field integral. The nominal values were converted into a current by the control program, using the slope as given by Eq. (4.6). In Fig. 4.26 also some details on diagnostics of the spectrometer are visible. Magnet 3B1 was equipped with two NMR probes and one Hall probe, whereas the other magnets had only one NMR probe. The standard current used was $\pm$ 150 A, which corresponds to a B-field of $\pm$ 0.11 T and a 5 mm dispersion in the center of the chicane (mid-chicane) for an electron beam of 28.5 GeV. The two NMR probes in 3B1 were of different types with different working ranges. They, however, overlapped in the region of the B-field setting. The Hall probe, placed in magnet 3B1 only for runs in March 2007, was removed after that to be installed in the wiggler magnet for experiment T475.

Along the beam line, two flux gate monitors were also placed in the mid-chicane to provide access to the stray field. One was placed on the girder to read the X- and Y-field components and the other on the beam pipe reading only the Y-component.

5 Relative Beam Energy Resolution

5.1 General Considerations

During 2007 a complete 4-magnet chicane was commissioned in ESA so that beam energy measurements could be performed. Here only some general arguments are discussed which are needed to evaluate the resolution of the measurements. The structure of this chapter is as follows. In Sect. 5.2 the BPMs within the SLAC A-line (the energy BPMs) will be described including their resolution. Section 5.3 evaluates the beam energy resolution of the 4-magnet chicane and compares the results with those from the energy BPMs. The last section, Sect. 5.4, contains the conclusions.

The data from the experiment T474/491 were collected in "runs", where a run had a duration of typically $10 \div 20$ minutes. The data were handled by a Labview program running on a Windows XP computer and, after some reorganization, the measurements were stored into root files.

For a given run, the field of the magnets was kept fixed and only changed periodically for positive polarity, zero current and negative polarity runs. Over the 3-weeks data taking period, several operation modes such as energy scans and calibration procedures were performed. Calibration runs also collected data from the corrector dipoles, Helmholtz coils and mover systems.

Energy scans were performed by changing the beam energy from the nominal value $E_b = 28.5$ GeV in 5 steps of 50 MeV between $E_b \pm 0.1$ GeV, i.e. over a 200 MeV total energy range. In the following sections only these data will be discussed.

5.2 Energy BPMs

Within the bend in the A-line, some BPMs were installed to provide the relative variation of E_b, the nominal beam energy, through their X-position readings. The data available were the X-position and the tilt in the XZ-plane of BPMs 12 and 24. In the following, these variables are denoted as x12Pos, x24Pos, x12Tilt and x24Tilt. At the locations of the BPMs in the A-line, the X-positions and the tilts θ of the beam are related to the beam energy through $E_b \propto 1/x \propto 1/\theta$, neglecting possible position jitters of the beam. Thus, the relative energy resolution σ_{E_b}/E_b can be estimated as

$$\frac{\sigma_{E_b}}{E_b} = \frac{\sigma_{\text{X}}}{\text{x}}, \qquad (5.1)$$

where x represents one of the variables x12Pos, x24Pos, x12Tilt and x24Tilt.

In a first step, all quantities used were normalized in order to be able to compare them to each other. Denoting x as a generic raw variable, the normalized quantity x' is defined as

$$x' = \frac{x - \bar{x}}{\Delta_x}, \qquad (5.2)$$

where $\bar{x}$ is the mean value of the variable x in question and Δ_x the standard deviation calculated for a given number of events.

In Figs. 5.1 and 5.2 position measurements of BPM 12 and 24 are shown from an energy scan performed for a run with the magnet current of +150 A. Each point corresponds to a beam bunch of 10^{10} particles and as seen on the figures, for a fixed energy a cluster of points exists which is mainly due to the beam energy jitter. The five energy steps of the scan at E_b-100 MeV, E_b-50 MeV, E_b, E_b+50 MeV, E_b+100 MeV are clearly visible.

In Fig. 5.1 the concept of the normalization procedure is addressed for the raw X-position data of BPM 12, shown left. Since the BPM values shown are not calibrated, the Y-axis is given in arbitrary units. The histogram on the right-hand side is the projection of the data of Fig. 5.1 (left) to the Y-axis, from which the mean value $\bar{x}$ and the RMS Δ_x can be deduced. It is important to note that Δ_x and the resolution σ_{X} in Eq. (5.1) are different quantities. The first quantity represents the width of the data used for normalization, while the latter one denotes the error on position or tilt measurements.

An example of the normalization is presented in Fig. 5.2, where for an energy scan the raw X-position readings of BPM 12 and 24 (left) and the normalized data (right) are shown. The raw x24Pos measurements have different mean values and standard deviation than those of BPM 12, but after normalization, both data sets superimpose almost perfectly. In plots of normalized data, the Y-axis will be always given in "a.u.", i.e. in arbitrary units. .

The two dashed lines on the right-hand side of Fig. 5.2 indicate the total range in energy where the scan was performed. This range of about 3 for the normalized case corresponds to 200 MeV in terms of beam energy. This correspondence of 3 to 200 MeV provides the scale factor of the energy resolution procedure.

5.2.1 Energy BPM Resolution

Considering the normalized variables x24Pos, x12Pos, x24Tilt, x12Tilt as independent, it is possible to estimate their resolutions by means of the histograms in Fig. 5.3. Each histogram represents, for a given energy scan, the difference between two out of the four variables, bunch-to-bunch. Not all possible combinations are shown. It is important to note that the difference

5.2 Energy BPMs

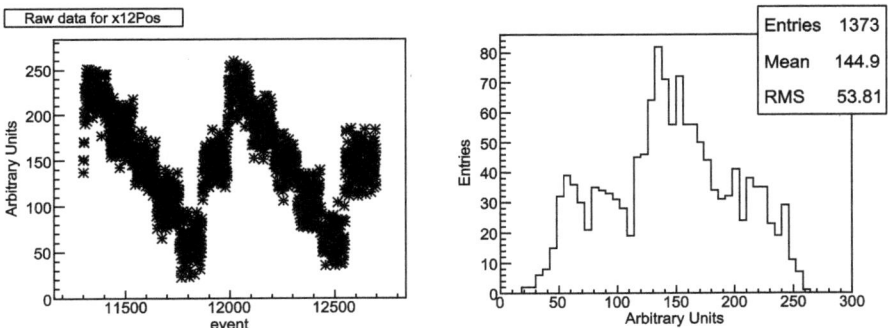

Figure 5.1: Left: Raw X-position readings of BPM 12 for an energy scan. Right: Projection of the data on the left-hand side to the Y-axis.

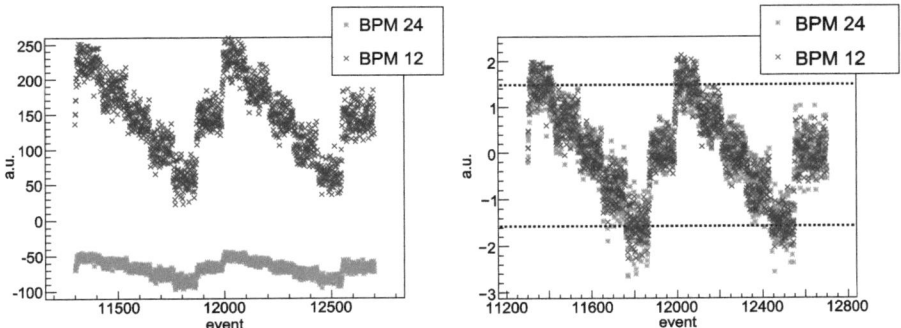

Figure 5.2: Left: Raw variables x12Pos and x24Pos for an energy scan. Right: Position data of the left-hand side after normalization. The two dashed lines indicate the scan range used.

between two quantities with different dimensions, like position and tilt, is now after normalization physically meaningful.

From Fig. 5.3 one derives

$$\begin{cases} \sigma_1^2 = \sigma_{x24Pos}^2 + \sigma_{x12Pos}^2 = (0.1865)^2 \\ \sigma_2^2 = \sigma_{x24Pos}^2 + \sigma_{x12Tilt}^2 = (0.2151)^2 \\ \sigma_3^2 = \sigma_{x12Pos}^2 + \sigma_{x24Tilt}^2 = (0.1231)^2 \\ \sigma_4^2 = \sigma_{x12Pos}^2 + \sigma_{x12Tilt}^2 = (0.125)^2 \,, \end{cases} \quad (5.3)$$

while, if the measurements are independent and possible position jitters are neglected, all σ's should be zero for an ideal situation. Any non-zero value indicates a finite resolution of the

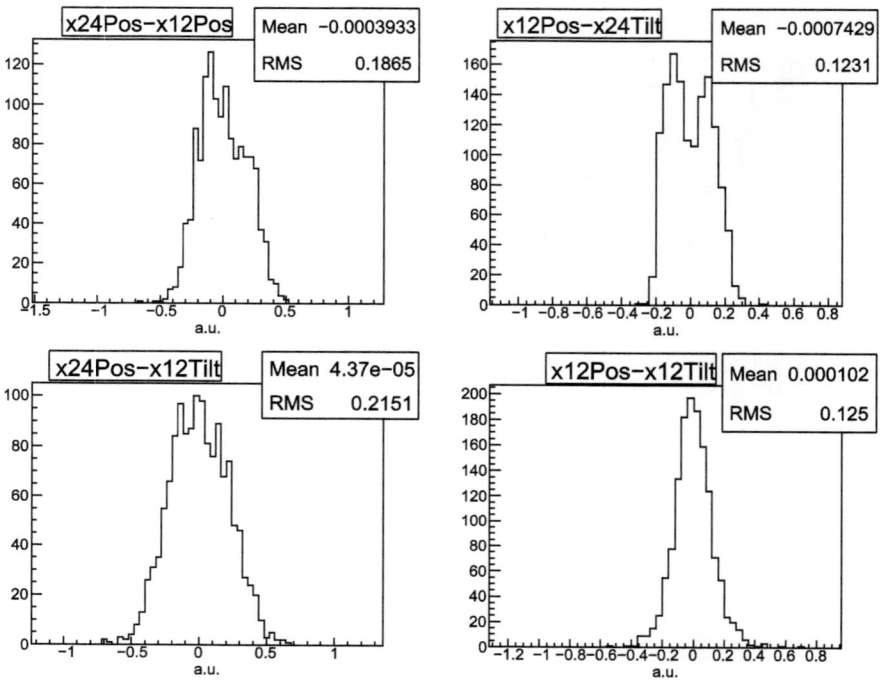

Figure 5.3: Differences between the variables as indicated for an energy scan.

BPM. Solving the system (5.3) for σ_{x24Pos}, σ_{x12Pos}, $\sigma_{x24Tilt}$ and $\sigma_{x12Tilt}$, we obtain

$$\begin{cases} \sigma^{a.u.}_{x24Pos} &= 0.187 \\ \sigma^{a.u.}_{x12Pos} &= 0.018 \\ \sigma^{a.u.}_{x24Tilt} &= 0.12 \\ \sigma^{a.u.}_{x12Tilt} &= 0.124 \, , \end{cases} \quad (5.4)$$

where $\sigma^{a.u.}$ means that the quantity in question is expressed in arbitrary units of the normalized scale. According to Eqs. (5.1) and (5.2), we obtain

$$\frac{\sigma^{a.u.}}{3} = \frac{\sigma^{MeV}}{200} \, , \quad (5.5)$$

so σ^{MeV}_{x12Pos}, defined as the resolution of the variable x12Pos for a relative beam energy measurement, can be written as $\sigma^{MeV}_{x12Pos} = 0.018 \cdot 200/3 = 1.2$ MeV, which corresponds to

$$\frac{\sigma^{MeV}_{x12Pos}}{E_b} = \frac{1.2 \text{ MeV}}{28.5 \text{ GeV}} = 4.2 \cdot 10^{-5} \, . \quad (5.6)$$

The other variables like x24Pos provide worse beam energy resolutions, up to an order of magnitude.

It is worthwhile to remind that the energy BPMs do not provide an absolute but only a relative beam energy measurement, i.e. no determination of E_b itself is possible.

5.3 ESA Magnetic Chicane

The basic scheme of the spectrometer in End Station A was shown in Figs. 3.5b and 4.26. Unfortunately, in the 2007 March runs the readout of the NMR probes was not working properly and in July the two NMR probes installed in magnet 3B1 and the one in 3B3 were damaged. Thus, essential informations for absolute beam energy measurements were missing. Moreover, since a complementary or independent method for E_b measurement was a priori not foreseen in ESA, the results from the chicane could not be cross-checked or cross-calibrated. Only relative beam energy measurements could be performed.

5.3.1 Mid-chicane BPM 4

For the following analysis only data from the July 2007 runs are taken into account and, if not explicitly stated, all variables considered are unnormalized. Since BPM 7, also positioned in the mid-chicane, was not reliable, only data from BPM 4 will be employed.

Selecting the energy scan data from run 2743 as an example, Fig. 5.4 shows the X-reading of BPM 4 (x4Pos). Here, steps in energy cannot be recognized, unlike to Fig. 5.2, since position fluctuations of the beam are bigger than the energy steps applied.

In fact, referring to Figs. 3.2 and 5.5, the dispersion d can be written as

$$d = x^{(4)}_{chicane}(E_b) = \text{x4Pos} + A - x^{(4)}_{jitter} \; , \tag{5.7}$$

where $x^{(4)}_{chicane}(E_b)$ is the beam offset in the mid-chicane at the BPM 4 beam line position, which is coupled to the beam energy via Eq. (2.19), $x^{(4)}_{jitter}$ the extrapolation position of the beam at the same Z-position (see Fig. 5.5) and A the BPM offset, which ensures to locate the

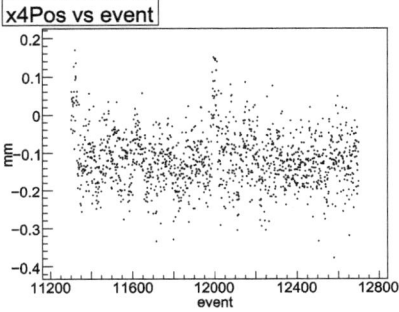

Figure 5.4: Beam positions of BPM 4 (without normalization) of an energy scan.

beam close to the center of the BPM. The value of A varies with the B-field, for example when switching the magnet polarity or when running with zero-magnet current. Nevertheless, since we normalize the data at the end of the analysis, this offset is not important and allows us to write

$$\text{x4Pos} = x^{(4)}_{chicane}(E_b) + x^{(4)}_{jitter} \ . \tag{5.8}$$

In order to have access to the offset $x^{(4)}_{chicane}(E_b)$ it is necessary to subtract $x^{(4)}_{jitter}$ from x4Pos. $x^{(4)}_{jitter}$ can be deduced from an extrapolation of the beam using position values from BPMs 1,2,3,5,9,10,11 (see Fig. 3.5b). It is obvious that for zero-current runs $x^{(4)}_{chicane}(E_b) = 0$ and, hence, x4Pos=$x^{(4)}_{jitter}$.

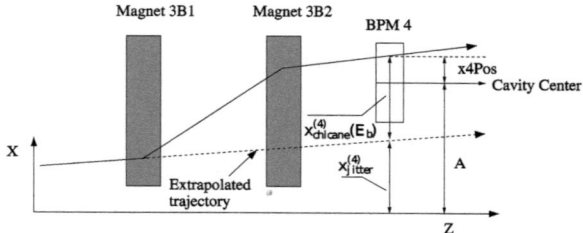

Figure 5.5: Working principle of the chicane: a charged particle traveling through the spectrometer receives an additional offset inversely proportional to its energy. The X-position reading of BPM 4 is the sum of two quantities, the extrapolated position $x^{(4)}_{jitter}$ and the offset $x^{(4)}_{chicane}(E_b)$.

5.3.2 Evaluation of $x^{(4)}_{jitter}$

The simplest way to determine $x^{(4)}_{jitter}$ is to perform a linear fit through the BPM beam position values upstream and downstream of the chicane and to extrapolate the straight line to the Z-position of BPM 4. This method is, however, limited since it does not take into account any coupling between X- and Y-positions of the BPMs and other informations as the tilt. Rotation or misalignment informations of the BPMs are missing and could therefore not be implement into the analysis. Also, the offsets of the center of the BPMs should be known, whose determination is also a non-trivial task.

Figure 5.6 shows $x^{(4)}_{jitter}$ as inferred from BPMs 1,2,3,5 upstream of the chicane against the BPM 4 X-position readings. The data are from a run, where the magnet current was set to zero, so that x4Pos = $x^{(4)}_{jitter}$. The dashed line in Fig. 5.6 represents a straight line with a slope of 1. It is evident that x4Pos = $\alpha \cdot x^{(4)}_{jitter}$, with $\alpha \neq 1$, which might be caused by e.g. different offsets or rotations of the BPMs around the Z-axis. In general, x4Pos can be written as

$$\text{x4Pos} = \alpha \cdot x^{(4)}_{jitter} + \beta \cdot x^{(4)}_{chicane}(E_b) \ , \tag{5.9}$$

5.3 ESA Magnetic Chicane

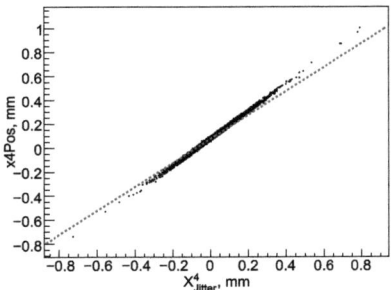

Figure 5.6: BPM 4 position measurement vs. $x_{jitter}^{(4)}$ evaluated from a linear extrapolation of the X-positions from BPMs 1,2,3,5.

where β denotes a possible relative rotation of BPM 4 with respect to the dipole field.

Thus, a more appropriate approach seems to be necessary to predict the correct value of $x_{jitter}^{(4)}$, with $\alpha = 1$ (a discussion on β will be presented later, but for the moment we assume $\beta = 1$). One option for an improved $x_{jitter}^{(4)}$ evaluation relies on the assumption that $x_{jitter}^{(4)}$ depends linearly on other BPM variables upstream and/or downstream of the chicane. Figure 5.7 shows, as an example, x4Pos plotted against x5Pos (left) and x5Tilt (right) from BPM 5. The data are not normalized and were taken from a run with zero-current (run 2747). A clear correlation between the variables exists in both cases. The beam jitter at BPM 4 can therefore be written as

$$\begin{aligned} x_{jitter}^{(4)} = &\ c^0 + c_1^1 \cdot \text{x1Pos} + c_1^2 \cdot \text{x1Tilt} + c_1^3 \cdot \text{y1Pos} + c_1^4 \cdot \text{y1Tilt} \\ &\ \vdots \\ &\ + c_j^1 \cdot \text{xjPos} + c_j^2 \cdot \text{xjTilt} + c_j^3 \cdot \text{yjPos} + c_j^4 \cdot \text{yjTilt} \\ &\ \vdots \\ &\ + c_N^1 \cdot \text{xNPos} + c_N^2 \cdot \text{xNTilt} + c_N^3 \cdot \text{yNPos} + c_N^4 \cdot \text{yNTilt}, \end{aligned} \quad (5.10)$$

where xjPos and yjPos denote the X-, respectively, Y-position, xjTilt and yjTilt the corresponding tilts of the j-th BPM and N the total number of BPMs used. For a run with magnet current set to zero, the coefficients c_i^j are determined by minimizing the quantity

$$\sum_{i=1}^{N_{event}} (\text{x4Pos} - x_{jitter}^{(4)})_i^2 \ . \quad (5.11)$$

The results for the coefficients c_i^j so obtained are discussed in Sect. 5.3.4.

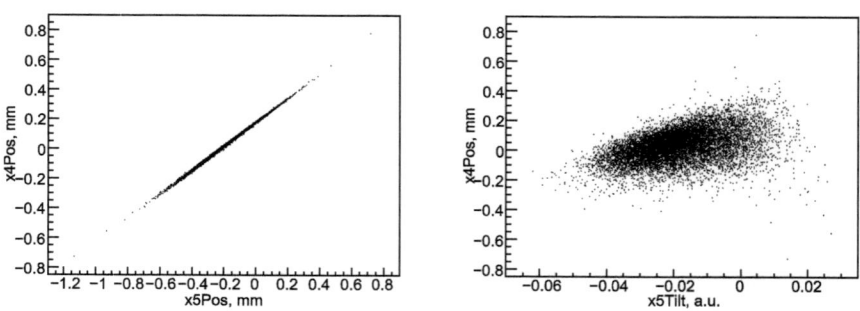

Figure 5.7: x4Pos as a function of x5Pos (left) and x5Tilt (right) for a run with zero-current.

5.3.3 Dipole Magnets

An essential prerequisite of the spectrometer is that the beam position downstream of the chicane is not coupled with that within the chicane, which means that the upstream beam path must be restored downstream. In other words, the chicane has to act on the beam in a symmetric manner. B-field measurements were performed in March 2007 to study the response of the chicane. Some results are shown in Fig. 5.8. Here, differences between the measured and nominal B-fields are plotted as a function of the nominal value, for negative and positive polarities.

For magnet 3B1, only the Hall probe was employed, whereas for the other magnets NMR probes were used. As can be seen, the differences vary from few tenths of mT (for magnets 3B2 and 3B3) up to about 3 mT. Magnet 3B4 shows field values much closer to the nominal ones, because only for this magnet the relation between the current and the field as given in Sect. 4.4.6 was determined and used for the field settings. The differences seen in the Fig. 5.8 might be attributed to any residual field, which was estimated to be $0.2 \div 0.4$ mT (see Sect. 4.4.1.3) and which is expected to depend on the history of the magnet and on the steel properties (no careful design and composition of the steel were accounted for). As a consequence, the path of the beam upstream could not fully restored downstream by the chicane, and changes of the beam energy are converted to position variations in BPMs 9, 10 and 11.

This supposition is supported by the following discussion. Similarly to what was done for BPM 4, the X-position of BPM 9 downstream of the chicane can be written as

$$\text{x9Pos} = x^{(9)}_{jitter} + x^{(9)}_{chicane}(E_b) \ . \tag{5.12}$$

For simplicity, α and β are assumed to be 1. If BPM 1, 2, 3 and 5 position data are used to predict the X-position jitter at BPM 9 ($x^{(9)}_{jitter}$) and if this value is subtracted from x9Pos for an energy scan, $x^{(9)}_{chicane}(E_b)$ as shown in Fig 5.9 is obtained. The steps due to different beam energies are basically visible which should however not be the case. Only position jitters with horizontal fluctuations around zero are expected. This observation supports that the BPMs

5.3 ESA Magnetic Chicane

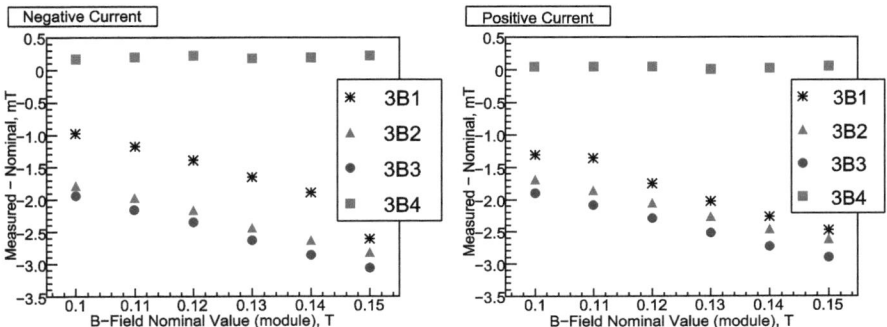

Figure 5.8: Differences between the measured and nominal B-fields as a function of the nominal value of the four magnets in ESA. Left: Negative current. Right: Positive current.

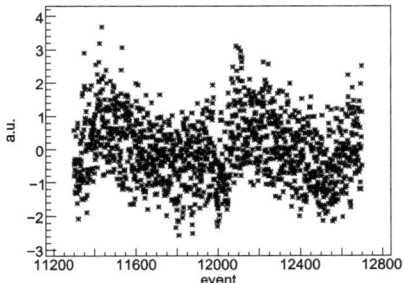

Figure 5.9: Normalized beam position of BPM 9 for an energy scan after subtraction of the jitter obtained from BPMs 1, 2, 3 and 5.

downstream of the chicane are not useful to be used to estimate the beam jitter in BPM 4, $x_{jitter}^{(4)}$.

5.3.4 Energy Resolution of the Spectrometer

Using data from a particular zero-current run, the parameters c_i^j were determined by minimizing expression (5.11), described in Sect. 5.3.2, by accounting for X- and Y-position as well as X- and Y-tilt data from BPMs 1, 2, 3 and 5.[1]

Using the coefficients c_i^j obtained, $x_{jitter}^{(4)}$ can be calculated from Eq. (5.10) and if x4Pos is now plotted against the new $x_{jitter}^{(4)}$, a straight line with a slope close to 1 is obtained, see Fig. 5.10. This result demonstrates that $x_{jitter}^{(4)}$ after minimizing (5.11) is much more suitable than the procedure discussed at the beginning of Sect. 5.3.2.

---

[1] If only X-position and X-tilt measurements are taken into account, less precise results are obtained, see also the discussions in Sect. 5.3.4.1.

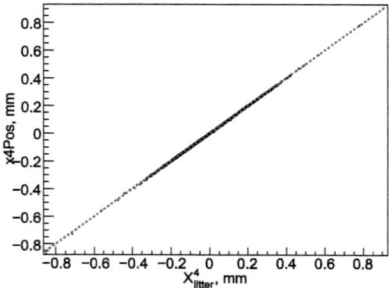

Figure 5.10: The position x4Pos versus $x_{jitter}^{(4)}$ evaluated using Eq. (5.10). The data are from a run with magnet current set to zero.

5.3.4.1 Beam Energy Resolution

The relative beam energy resolution of the 4-magnet chicane was performed in the following way. Considering an energy scan with magnet current of +150 A and enabling subtraction of $x_{jitter}^{(4)}$ from x4Pos, the beam offset in the mid-chicane, $x_{chicane}^{(4)}(E_b)$, is obtained from Eq. (5.8). In Fig. 5.11 (left) the normalized x12Pos and $x_{chicane}^{(4)}(E_b)$ data are superimposed, while on the right-hand side their difference (x12Pos - $x_{chicane}^{(4)}(E_b)$) is shown. The non-zero standard deviation of the histogram given in arbitrary units, $\sigma^{a.u.}$, is due to the finite resolution of $x_{chicane}^{(4)}(E_b)$ as well as of x12Pos. In fact,

$$\sigma^{a.u.} = \sigma^{a.u.}_{x_{chicane}^{(4)}(E_b)} \oplus \sigma^{a.u.}_{x12Pos} = 0.365 \;, \tag{5.13}$$

where the two terms were added in quadrature. $\sigma^{a.u.}_{x12Pos}$ was evaluated as 0.0187 in Sect. 5.2.1 and is negligible, so that

$$\sigma^{a.u.} \simeq \sigma^{a.u.}_{x_{chicane}^{(4)}(E_b)} \simeq 0.365 \;. \tag{5.14}$$

$\sigma^{a.u.}_{x_{chicane}^{(4)}(E_b)}$ can be understood as the normalized beam energy resolution of the 4-magnet chicane. Accounting for the correspondence between normalized and unnormalized variables, we conclude

$$\sigma^{MeV}_{x_{chicane}^{(4)}(E_b)} = \sigma_{E_b} = 0.365 \cdot 200/3 = 24.3 \text{ MeV}$$
$$\rightarrow \frac{\sigma_{E_b}}{E_b} = \frac{24.3 \text{ MeV}}{28.5 \text{ GeV}} = 8.5 \cdot 10^{-4} \;. \tag{5.15}$$

If only the X-positions and X-tilts of BPMs 1, 2, 3 and 5 are taken into account, $x_{jitter}^{(4)}$ is less precise and the energy resolution becomes $\sigma_{E_b}/E_b \simeq 8.9 \cdot 10^{-4}$. This value is close to that in Eq. (5.15), which means that the variables yPos and yTilt have no strong impact on $x_{jitter}^{(4)}$ and, hence, on the energy resolution.

5.3 ESA Magnetic Chicane

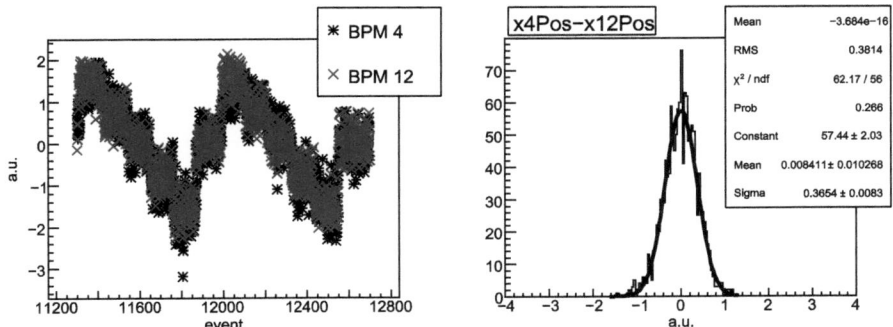

Figure 5.11: Left: Normalized position x4Pos after subtraction of the jitter and the position x12Pos, for an energy scan with magnet current of +150 A. Right: Difference between $x_{chicane}^{(4)}(E_b)$ and x12Pos.

In principle, σ_{E_b}/E_b can be estimated in a complementary way. Since $x_{chicane}^{(4)}(E_b)$ = x4Pos - $x_{jitter}^{(4)}$, the resolution of $x_{chicane}^{(4)}(E_b)$ is equivalent to the resolution of the difference (x4Pos - $x_{jitter}^{(4)}$). Figure 5.12 shows this difference for a run with zero magnet current. For this case, (x4Pos - $x_{jitter}^{(4)}$) is expected to be zero for every bunch, but the resulting standard deviation is non-zero. Interpreting this non-zero value as the resolution of (x4Pos - $x_{jitter}^{(4)}$), respectively, $x_{chicane}^{(4)}(E_b)$ and accounting for the dispersion $x_{chicane}^{(4)}(E_b) \simeq 5$ mm in the mid-chicane, Eq. (5.1) becomes

$$\frac{\sigma_{x_{chicane}^{(4)}(E_b)}}{x_{chicane}^{(4)}(E_b)} = \frac{\sigma_{E_b}}{E_b} = \frac{2.3\ \mu m}{5\ mm} = 4.6 \cdot 10^{-4}\ . \quad (5.16)$$

This result is considerably better than that in (5.15). Possible reasons for the difference might be some non-demagnetization after switching off the magnets and/or a non-equal to 1 value of β in Eq. (5.9). In fact, if a rotation of BPM 4 with respect to the dipole field by an angle θ exists, the effective offset at the mid-chicane is 5 mm $\cdot \cos\theta$ and the error on the relative beam energy measurement becomes

$$\frac{\sigma_{E_b}}{E_b} = \frac{2.3\ \mu m}{5\ mm \cdot \cos\theta} > 4.6 \cdot 10^{-4}\ , \quad (5.17)$$

which points into the right direction.

5.3.5 X- and Y-Position Coupling

In analogy to X-jitter determination, it is possible to estimate the jitter in Y-direction, $y_{jitter}^{(4)}$, and, after subtraction from y4Pos, $y_{chicane}^{(4)}(E_b)$ can be deduced. In Fig. 5.13, $y_{chicane}^{(4)}(E_b)$ is plotted against $x_{chicane}^{(4)}(E_b)$ for an energy scan. Some correlation between X- and Y-positions is clearly visible, which reveals that β in Eq. (5.9) is definitely different from 1. This can be caused by a rotation of BPM 4 relative to the magnetic field as already stated in the previous

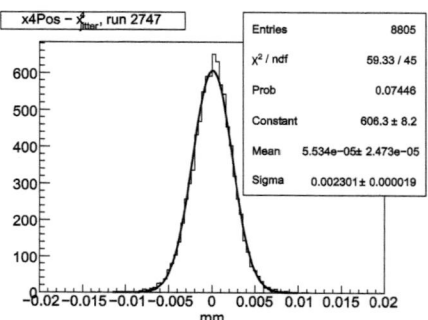

Figure 5.12: Difference between x4Pos and $x^{(4)}_{jitter}$ for a zero-current run. $x^{(4)}_{jitter}$ is calculated using the parameters from the minimization procedure as described in Sect. 5.3.2.

section. Thus, to improve the relative beam energy resolution, also $y^{(4)}_{chicane}(E_b)$ should be taken into account in the analysis.

One way to include $y^{(4)}_{chicane}(E_b)$ is to perform a rotation in the (X,Y)-plane. After jitter subtraction, two new variables are defined as

$$\begin{pmatrix} (x^{(4)}_{chicane}(E_b))_{rot} \\ (y^{(4)}_{chicane}(E_b))_{rot} \end{pmatrix} = \begin{pmatrix} \cos\theta & -\sin\theta \\ \sin\theta & \cos\theta \end{pmatrix} \begin{pmatrix} x^{(4)}_{chicane}(E_b) \\ y^{(4)}_{chicane}(E_b) \end{pmatrix}. \qquad (5.18)$$

The rotation angle $\theta = \arctan(-0.4374)$ equals to the angle of the straight line in Fig. 5.13. However, after normalization and comparison of $(x^{(4)}_{chicane}(E_b))_{rot}$ with x12Pos, no improvement for σ_{E_b}/E_b was found when compared with the values in the previous section. One reason could be the less precise resolution of the difference (y4Pos - y^4_{jitter}), which was found to be 3.2 μm compared to 2.3 μm (see Eq. (5.16) and Fig. 5.12). Furthermore, the value of $\theta = \arctan(-0.4374) = 23.49°$ cannot convincingly alone explain the difference between the two resolution values. Indeed, the amount of rotation needed to remove the difference is

$$\frac{\sigma_{E_b}}{E_b} = \frac{2.3\ \mu m}{5\ mm \cdot \cos\theta} = 8.5 \cdot 10^{-4}$$
$$\rightarrow \theta = \arccos\left(\frac{2.3\ \mu m}{5\ mm \cdot 8.5 \cdot 10^{-4}}\right) = 57°, \qquad (5.19)$$

which is much too large when compared to the expectation of at most some degrees.

5.4 Summary

The beam energy resolution, σ_{E_b}/E_b, was measured by employing a 4-magnet chicane in End Station A at the Stanford Linear Accelerator Center. The chicane was commissioned in 2006/2007 for the experiment T474/491. The resolution σ_{E_b} is understood as the smallest variation of energy which can be measured in a reliable way by the spectrometer. Some

5.4 Summary

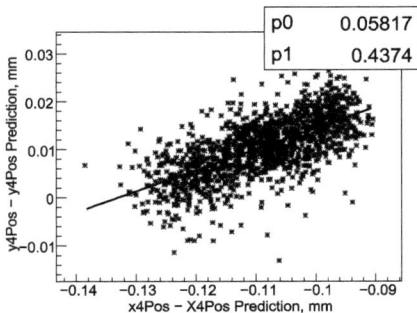

Figure 5.13: $y^{(4)}_{chicane}(E_b)$ versus $x^{(4)}_{chicane}(E_b)$ for an energy scan.

imperfect readings of the NMR probes in the magnets, missing information on BPM alignment and, more important, the lack of a complementary or redundant method for absolute E_b measurements allowed only to perform relative beam energy determinations.

Due to large beam position jitters in the mid-chicane, a method was developed to subtract this jitter using data from BPMs outside the chicane. The method is in general based on the assumption that the jitter of interest can be written as a linear combination of X- and Y-positions as well as X- and Y-tilts from the BPMs upstream and downstream of the chicane.

In order to employ the BPMs downstream of the chicane it has to be ensured that the upstream beam path is fully restored downstream of the chicane. Unfortunately, this condition was not realized, probably caused by some non-demagnetization effects and different responses of the magnets using a common current. Such shortcomings are able to introduce a beam offset in the BPMs downstream, which is correlated with the beam energy. Thereby, these BPMs could not be used for beam jitter measurements and, as a consequence, some less precise beam energy resolution is obtained.

In summary, the beam energy resolution of the 4-magnet chicane, σ_{E_b}/E_b, has been measured as

$$\frac{\sigma^{MeV}_{x4Pos}}{E_b} = \frac{\sigma_{E_b}}{E_b} = \frac{24.3 \text{ MeV}}{28.5 \text{ GeV}} = 8.5 \cdot 10^{-4} \ . \tag{5.20}$$

This value is larger than the request of 10^{-4} (or better) for the ILC, but substantial improvements for future BPM-based spectrometers, as discussed in the thesis, indicate that the ILC goal can be achieved.

6 Laser Compton Energy Spectrometer

6.1 General Considerations

To fully exploit the physics at the International Linear Collider, excellent control of the beam energy is mandatory. The device for the best candidate of this purpose is the 4-magnet chicane spectrometer of which a prototype was discussed in Chapter 3.

In Chapter 2 an overview of concepts for beam energy measurements was given, including some experiences derived from spectrometers which were in operation. This part of the thesis also emphasizes the importance of more than one technique to be implemented in order to permit cross-calibration and cross-checks of precise $\sqrt{s}$ determinations. In the past, novel suggestions for beam energy measurements were proposed, see e.g. [51], which should operate independently from the magnetic chicane.

In this Chapter we propose a novel, non-destructive method for beam energy measurements using Compton backscattering of laser light off beam particles. This method is intended to be used as a complementary determination of the beam energy in conjunction with the BPM-chicane. An introduction of the physics of the Compton process was given in Sect. 2.1.2, and some applications of this process for beam energy measurements in the past were described in Sects. 2.2.1 and 2.2.2, verifying its feasibility and excellent performance.

Unfortunately, the method described in these sections is, however, not practicable for the ILC, since for each bunch crossing accumulation of large statistics is necessary, which precludes energy measurements of single photons. Furthermore, a precise calibration of the corresponding calorimeter is also not possible as it was realized at BESSY and VEPP-4M, using radioactive sources of nearly monochromatic photons in the energy range of the backscattered photons. At the ILC, the photons are in the GeV range where precise and fast calibration is very challenging or even excluded.

Therefore, the method proposed for the linear collider has to be different and can be summarized as follows. After crossing the electron beam with laser light, a dipole magnet separates three types of particles: the Compton scattered photons and electrons as well the non-interacting beam particles. Downstream of the magnet, detectors record the position of the backscattered photons and of those electrons having smallest energy or largest deflection. If these measurements are combined either with informations on the integrated B-field of the magnet or with the unscattered beam particle position, the beam energy can be deduced.

Chapter 6 is organized as follows. Section 6.2 reviews physics aspects of the Compton process not presented in Sect. 2.1.2. In particular, aspects which are critical for our method are discussed. In Sect. 6.3 the general layout of the spectrometer is presented and two novel approaches for beam energy determination are discussed in some details in order to understand their advantages and disadvantages and the precisions achievable for each method. In Sect. 6.4 detector options for the Compton photons and electrons are discussed and simulation will be presented to demonstrate the feasibility and reliability of the concepts proposed. Section 6.5 describes the basic requirements of the laser system needed. In Sect. 6.6 processes beyond the Born approximation within the laser-electron interaction region such as non-linear effects, multiple scattering, higher order QED contributions and the Breit-Wheeler pair production background are discussed and their impact on the measurements is evaluated. This will be followed by a general discussion and estimations of additional potential systematic errors (Sect. 6.7). Possible locations for the energy spectrometer are summarized in Sect. 6.8 and, finally, conclusions are given in Sect. 6.9.

6.2 The Compton Scattering Process

In Fig. 6.1 the kinematics of the Compton scattering process in the laboratory frame is shown. Beam particles with energy E_b collide nearly head-on off laser photons with energy E_λ, producing scattered electrons and photons of energy E_e and E_γ, respectively. α is the collision angle between the laser and electron beam, whereas θ_γ is the scattering angle of the outgoing photons. The system of reference used throughout this study is, according to Fig. 6.1, defined in the following way: the Z-axis corresponds to the direction of the beam particles, the X-axis lies in the horizontal or bending plane and the Y-axis points to the vertical direction such that a right-handed coordinate system is obtained.

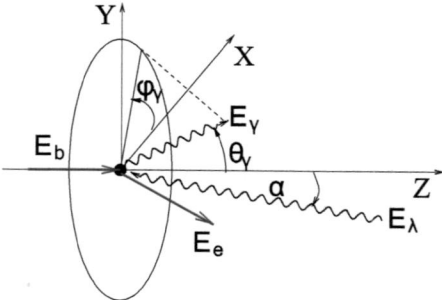

Figure 6.1: Kinematics of the Compton process in the lab frame. The incoming beam with energy E_b collides quasi head-on with laser photons of energy E_λ, producing scattered electrons and photons of energy E_e and E_γ, respectively.

6.2 The Compton Scattering Process

6.2.1 Compton Cross-Section

In Sect. 2.1.2 the differential cross-section for unpolarized electrons was presented, Eq. (2.16). The differential spin-dependent cross-section of the Compton process, obtained after summing over all possible spin states of the final state particles, is in Born approximation:

$$\frac{d\sigma}{dy} = \frac{2\sigma_0}{x}\left[\frac{1}{1-y} + 1 - y - 4r(1-r) + P_e\lambda rx(1-2r)(2-y)\right], \quad (6.1)$$

where y is the normalized energy variable, $y = 1 - E_e/E_b = E_\gamma/E_b$, x a dimensionless variable defined as

$$x = \frac{4E_b E_\lambda}{m^2} \cdot \cos^2 \alpha/2, \quad (6.2)$$

P_e the initial electron helicity ($-1 \leq P_e \leq 1$), λ the initial photon helicity ($-1 \leq \lambda \leq 1$), $r = y/[x(1-y)]$ and $\sigma_0 = \pi r_0^2 = 0.2495$ barn, with r_0 the classical electron radius.

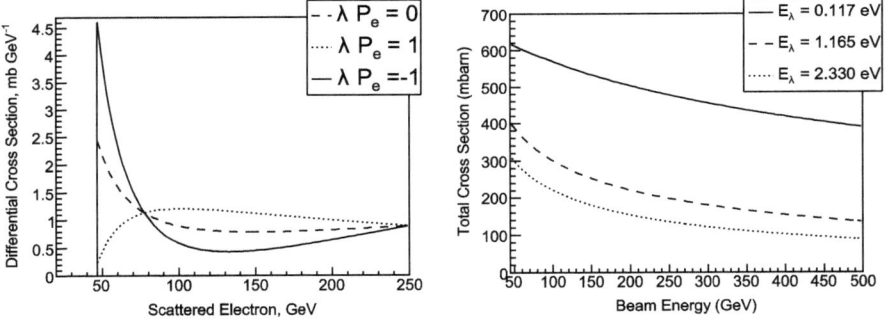

Figure 6.2: Left: Compton backscattering cross-section for three polarization configurations versus scattered electron energy for an infrared Nd:YAG laser at 250 GeV. Right: Compton backscattering cross-section versus beam energy for three laser energies and unpolarized beam electrons.

The right-hand side of Fig. 6.2 shows the unpolarized Compton cross-section as a function of the beam energy for three laser energies, $E_\lambda = 0.117$, 1.165 and 2.33 eV. At all incident energies, the CO_2 laser with $E_\lambda = 0.117$ eV provides the largest cross-sections, while for the Nd:YAG laser (with $E_\lambda = 1.165$ or 2.33 eV) the cross-sections are significantly smaller. For example, at 250 GeV the CO_2 cross-section is more than two times larger than the Nd:YAG laser values. We also note that for the polarization configuration $P_e\lambda = -1$, the cross-section close to the electron's kinematic endpoint is enhanced by typically a factor two, while for the configuration $P_e\lambda = +1$, the edge Compton cross-section vanishes. This behavior is shown in Fig. 6.2 (left), where for the three cases, $P_e\lambda = -1$, $P_e\lambda = +1$ and unpolarized, the cross-section is plotted as a function of the Compton electron energy for the infrared Nd:YAG laser at 250 GeV. For polarized electrons, the favored spin configuration $P_e\lambda = -1$ can always be achieved by adjusting the laser helicity λ. In this first part of the chapter, we consider fully polarized electrons and laser photons. In Sect. 6.4.2.1, possible impact of not fully polarized beam electron or laser on beam energy measurement will be briefly discussed.

6.2.2 Properties of the Final State Particles

Eq. (2.13) in Sect. 2.1.2 is calculated as a function of the photon scattering angle given in the electron rest frame for $\alpha = 0$. Considering a laser crossing the beam with an angle α, the angles of the scattered photons and electrons relative to the incoming beam direction are [52, 53]

$$\theta_\gamma = \frac{m}{E_b} \cdot \sqrt{\frac{x}{y} - (x+1)} \;, \qquad \theta_e = \theta_\gamma \cdot \frac{y}{1-y} \;, \tag{6.3}$$

and E_γ as a function of θ_γ in the lab frame and α can be written as

$$E_\gamma = E_\lambda \cdot \frac{1 + \beta \cos \alpha}{1 - \beta \cos \theta_\gamma + \frac{E_\lambda (1 + \cos(\theta_\gamma + \alpha))}{E_b}} \;, \tag{6.4}$$

with β the beam electron velocity divided by the speed of light, and α the angle between the laser light and the incident beam. E_γ ranges from zero to some maximum value

$$E_{\gamma,max} = \frac{E_b^2}{E_b + \frac{m^2}{4\omega_0}} \;, \qquad \omega_0 = E_\lambda \cdot \cos^2(\alpha/2) \;, \tag{6.5}$$

with m the electron mass. Eq. (6.5) is equivalent to Eq. (2.15) for $\alpha = 0$ and $\beta \to 1$.

Figure 6.3 illustrates the energy and X-position of the scattered photons at a plane located 50 m downstream of the Compton IP for three laser energies, $\alpha = 8$ mrad and $E_b = 250$ GeV. According to Eqs. (6.3) and (6.5), γ-rays with highest energy travel exactly forward.

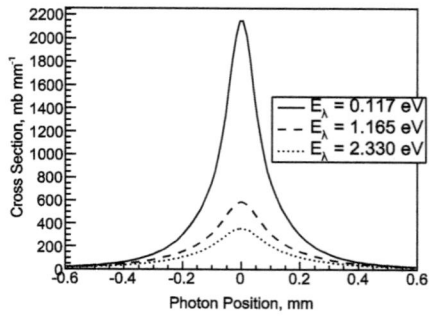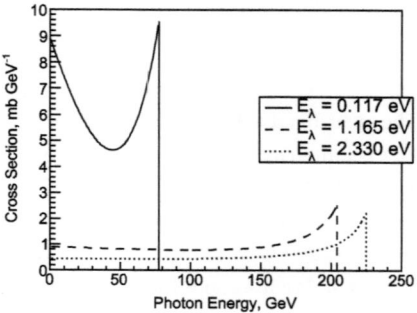

Figure 6.3: X-position (left) and energy (right) spectra for backscattered photons for three laser energies, $\alpha = 8$ mrad and $E_b = 250$ GeV. The photon position is determined at a plane 50 m downstream of the Compton IP. In both figures unpolarized electrons are assumed.

6.2 The Compton Scattering Process

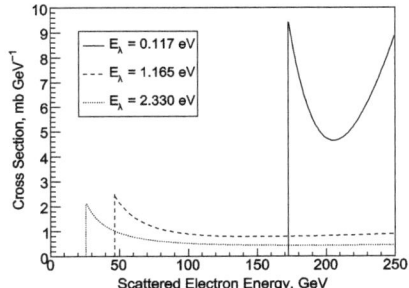

Figure 6.4: Unpolarized Compton cross-section versus scattered electron energy for three laser wavelengths at 250 GeV.

The energy of the Compton electrons is determined by energy conservation. The maximum energy of the Compton photon is related to the minimum (or edge) energy of the scattered electron, E_{edge}, via

$$E_{edge} = E_b + E_\lambda - E_{\gamma,max} = \frac{E_b}{1 + \frac{4E_b\omega_0}{m^2}} , \qquad (6.6)$$

if the laser energy E_λ is neglected. The electron scattering angle θ_e, given in Eq. (6.3), approaches zero as θ_γ becomes smaller. Thus, in the region of smallest electron energy, the region of our interest, both the scattered electrons and photons are generated at very small angles.

After the substitution $E_e = E_b - E_\gamma$, the unpolarized Compton cross-section as a function of the scattered electron energy for three laser energies at 250 GeV is obtained from Fig. 6.3 (right). The spectrum has an inverted shape, while the range is now $[E_{edge}, E_b]$ as seen in Fig. 6.4.

The CO_2 laser (with an energy of 0.117 eV) provides the most pronounced edge cross-section, while the Nd:YAG laser (with $E_\lambda = 1.165$ or 2.33 eV) cross-sections are significantly smaller. At the electron's edge position, both Nd:YAG lasers have cross-sections of similar size, with edge energy values relatively close to each other.

Since one of the proposed methods for measuring the beam energy utilizes the variation of the edge energy on E_b, see Eq. (6.6), we present in Fig. 6.5 the edge energy dependence on E_b for three laser wavelengths. As can be seen, the derivative dE_{edge}/dE_e or the sensitivity of the edge energy on E_b decreases with increasing laser energy. In particular for an infrared or a green laser, the sensitivity is very small, which suggests to employ lasers with large wavelengths, such as a CO_2 laser, for this method.

From these discussions we can draw the first conclusions that are relevant for the beam energy determinations:

- the electron edge energy, E_{edge}, depends on the beam energy (Eq. (6.6)), on which one of the proposals for measuring E_b relies;

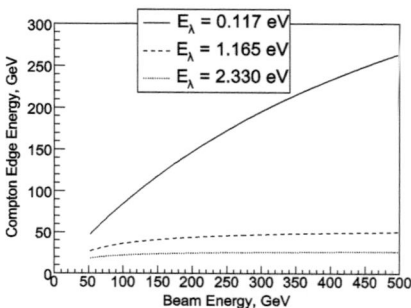

Figure 6.5: Edge energy of Compton backscattered electrons as a function of E_b for a CO_2, infrared and green laser.

- if this method will be utilized, low energy lasers are advantageous because of large Compton cross-section and high endpoint E_b sensitivity;

- backscattered electrons and photons are predominantly scattered in the direction of the incoming beam;

- photons associated with the edge electrons have largest energy and point toward $\theta_\gamma = 0$;

- the unpolarized Compton cross-section peaks at E_{edge}, which results in beam energy determinations with small statistical errors;

- for polarized electrons, choose the polarization configuration $P_e\lambda = -1$; the unfavored configuration $P_e\lambda = +1$ spoils E_b determination.

So far, the cross-section formulae and backscattered particle properties were discussed in the Born approximation. Possible modifications due to multiple scattering, e^+e^- pair background, higher order corrections and nonlinear effects were partially discussed in [54] and are further studied in Sect. 6.6.

6.3 Overview of the Energy Spectrometer

6.3.1 General Layout

Within the so-called single-event regime, individual Compton events originate from separate accelerator bunches. As was realized in experiments at storage rings [28, 29, 31], see also Sects. 2.2.1 and 2.2.2, recording the maximum energy of the scattered photons enabled to determine the beam energy.

The experimental conditions at the ILC with large bunch crossing frequencies and high particle intensity require to operate with short and intense laser pulses so that high instantaneous

6.3 Overview of the Energy Spectrometer

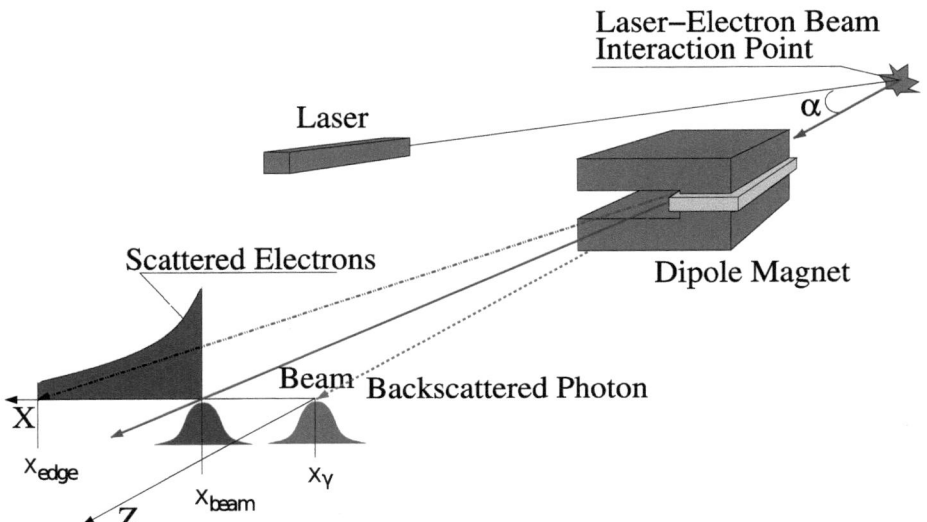

Figure 6.6: A scheme of the proposed energy spectrometer based on Compton backscattering.

event rates are achieved. As a result, the detector signals for a particular bunch crossing correspond to a superposition of multiple events. In such a regime, the measurement of the energy of a single photon cannot be realized and the signal will be an energy weighted integral over the photon spectrum covered by the angular acceptance of the detector. The number of Compton interactions should, however, be adjusted such that neither the incident electron beam will be disrupted nor the Compton event rate degrades the performance of the detectors. Discussions on possible detector solutions will be given in Sect. 6.4.

The concept of a possible Compton energy spectrometer is shown in Fig. 6.6. Downstream of the laser crossing point, a bending magnet is positioned which is followed by a dedicated particle detection system. This system has to provide precise position information of the backscattered photons and electrons close to the edge and, employing an alternative method, the position of the unscattered beam.

The vacuum chamber between the Compton IP and the detector plane needs some special design to accommodate simultaneously the trajectories of the photons, the deflected backscattered electrons and the non-interacting beam particles. In order to ensure large luminosity, the crossing angle should be very small and to protect optical elements from synchrotron radiation (present mainly in the horizontal plane) a vertical beam crossing is suggested.

To maximize the $e\gamma$ luminosity, the crossing angle α should be small, in our case $8 \div 10$ mrad, and the laser spot should be larger than the horizontal electron beam size, which is expected to be in the range of $10 \div 50$ μm within the beam delivery system[1]. For a well aligned laser

---

[1] The vertical beam size is much smaller and will not exceed few micrometers, resulting in horizontal/vertical aspect ratio of typically $10 \div 50$ within the BDS of the ILC.

it should be practicable to keep possible horizontal and vertical relative displacements of the electron and laser beams small enough, so that permanent overlap is ensured even in cases of beam position jitter.

The dipole magnet located about 3 m downstream of the crossing point separates the particles coming from the IP into the undeflected backscattered photons, the Compton electrons and the beam particles with smallest bending angle. The B-field integral should be scaled to the primary beam energy, so that beam particle deflection occurs always at the same angle (typically $0.5 \div 1$ mrad). Thus, one BPM with fixed position is sufficient to record the beam line position at all energies. The photon detector is located in the direction of the original beam, while the electron detector has to be adjusted horizontally according to Compton scattering kinematics and the magnetic field.

At ILC energies, Compton scattering with typical continuous lasers in the $1 \div 10$ Watt range takes some fraction of an hour to collect enough statistics for precise E_b determination. Thus, in order to perform bunch-to-bunch energy measurements the default laser system should be a pulsed laser with a pattern that matches the specific pulse and bunch structure of the ILC, i.e. at 250 GeV an inter-bunch spacing of ~300 ns within 1 ms long pulse trains at 5 Hz. In order to collect typically 10^6 Compton events per bunch crossing[2], the pulse power of the CO_2 laser should be about 1 mJ [3], while for an infrared laser with $E_\lambda = 1.165$ eV, the smaller Compton cross-section will be compensated by a smaller spot size and a power of 30 mJ. A laser in the green wavelength range with 2.33 eV photon energy requires a pulse power of 24 mJ for 10^6 Compton interactions. For Z-pole running, the laser power can be somewhat smaller (since the Compton cross-section is higher), but it has to be increased for upgraded 1 TeV runs. Since at present lasers with such properties are not commercially available, R&D is needed to achieve the objectives, see e.g. [55, 56].

The choice of a suitable laser system is determined by several constraints. Basically, lasers with large wavelengths such as a CO_2 laser provide high event rates due to large Compton cross-sections and best beam energy sensitivity of the endpoint position (see Fig. 6.5). Lasers in the infrared region such as Nd:YAG or Nd:YLF lasers, however, provide at present a better reliability, in particular with respect to the bunch pattern and pulse power [56] and would relax geometrical constraints of the spectrometer setup due to substantially smaller electron edge energies (see Fig. 6.4). Green laser R&D is ongoing within the ILC community to develop laser-wire diagnostics [57] and high energy polarimeters [52].

Figure 6.7 shows for three wavelengths and a particular setup (beam bending angle of 1 mrad and a detector 25 m downstream of the magnet) the horizontal or X-position of the Compton electrons at $E_b = 250$ GeV. The position of electrons with highest energy coincides with the beam line position independent of the laser, whereas the positions of the edge electrons with largest deflection are very distinct. Edge positions are smaller for larger laser wavelengths.

---

[2] 10^6 Compton events are sufficient to perform bunch-to-bunch measurements, as it will be shown in Sects. 6.3.2 and 6.3.3

[3] The laser power estimation assumes electron and laser beam parameters as discussed in Sect. 6.5

6.3 Overview of the Energy Spectrometer

For a CO_2 laser at $E_b = 45.6$ GeV, the edge electrons are separated by only 2.2 mm from the beam line, while they are displaced from the backscattered γ-rays by about 2.6 cm. Such space conditions would prevent the use of a CO_2 laser for Z-pole calibration runs. An increased B-field and/or a larger drift distance could somewhat relax the situation.

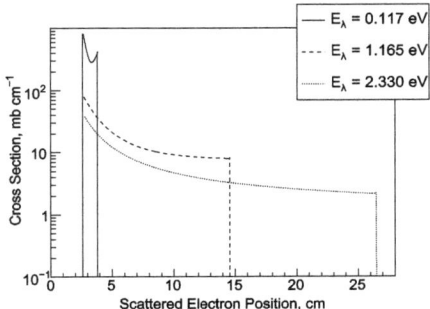

Figure 6.7: Scattered electron positions for $E_b = 250$ GeV, a B-field of 0.28 T and three laser energies. The detector is placed 25 m downstream of the spectrometer magnet. Since the transformation $X \propto 1/E$, the peak height of the Compton edge is reduced. For an infrared or a green laser it is not even visible anymore.

Lasers in the green or infrared wavelength region have some disadvantages. As already pointed out, they provide smaller Compton cross-sections and hence smaller event rates, which might only be compensated by higher laser power and/or smaller but limited spot sizes. Also, the smaller sensitivity of the edge position on E_b (Fig. 6.5) and the generation of additional background at large $\sqrt{s}$ due to e^+e^- pairs from Breit-Wheeler processes[4] might disfavor their application. As soon as the variable x of Eq. (6.2) exceeds 4.83, which is for example the case at 250 GeV and a green laser, e^+e^- pair production is kinematically possible[5]. Whether this source of background can be tolerated will be studied in Sect. 6.6. Some of the disadvantages discussed are of less relevance if an alternative method, method B, will be employed for beam energy determination.

In Sects. 6.3.2 and 6.3.3 two possible options of the layout described here will be illustrated. The first option, denoted as method A, is based on the determination of the electron edge energy measuring the distance $X_{edge} - X_\gamma$ (see Fig. 6.6), the B-field integral and the magnet-detector distance. From the value of E_{edge} and using Eq. (6.6) the beam energy can be calculated. The second method, method B, relies on measuring X_{edge}, X_{beam} and X_γ, from which it is possible to calculate the beam energy.

---

[4] These are $\gamma - \gamma$ interactions, where one γ stems from the Compton process and the other from the laser.
[5] The threshold of e^+e^- pair creation is $E_m E_\lambda = m^2 c^4$, with $E_m = x \cdot E_b/(x+1)$, which gives $x = 2(1 + \sqrt{2}) \simeq 4.83$.

6.3.2 Method A

One approach to measure the ILC beam energy by Compton backscattering relies on precise electron detection at the kinematic endpoint. In particular, endpoint or edge energy measurements are performed, from which via Eq. (6.6), the beam energy is accessible. In particular, the Compton edge electrons are momentum analyzed by utilizing the dipole magnet and recording their displacement downstream of the magnet.

The conceptual detector design consists of a component to measure the center-of-gravity of the Compton backscattered γ-rays[6] and a second one to measure the position of the edge electrons. The distance D of the center-of-gravity to the edge position and the well known drift space L between the dipole and the detector determine the bending angle Θ of the edge electrons, which, together with the B-field integral, fixes the energy of the edge electrons:

$$E = \frac{c \cdot e}{\Theta} \int_{magnet} Bdl \simeq \frac{c \cdot e \cdot L}{D} \int_{magnet} Bdl \,. \tag{6.7}$$

Here, c is the speed of light and e the charge of the particles[7]. Thus, for sufficient large drift space the edge electrons are well separated from the Compton scattered photons which pass the magnet undeflected.

A demanding aspect of this approach is the precision for the displacement, ΔD, which is related to the beam energy uncertainty as

$$\frac{\Delta E_b}{E_b} = \left(1 + \frac{4\omega_0 E_b}{m^2}\right) \sqrt{\left(\frac{\Delta B}{B}\right)^2 + \left(\frac{\Delta L}{L}\right)^2 + \left(\frac{\Delta D}{D}\right)^2} \,. \tag{6.8}$$

This relation follows from Eqs. (6.5), (6.6) and

$$\frac{\Delta E_{edge}}{E_{edge}} = \frac{E_{edge}}{E_b} \cdot \frac{\Delta E_b}{E_b} \tag{6.9}$$

as well as

$$\left(\frac{\Delta E_{edge}}{E_{edge}}\right)^2 = \left(\frac{\Delta B}{B}\right)^2 + \left(\frac{\Delta L}{L}\right)^2 + \left(\frac{\Delta D}{D}\right)^2 \tag{6.10}$$

together with $D = \Theta \cdot L$ from the geometry of the setup. Synchrotron radiation effects on $\Delta E_b/E_b$, estimated to be significantly smaller than any term in (6.8), were omitted.

One notices from Eq. (6.8) that smallest beam energy uncertainties are achievable for lasers with large wavelengths, such as a CO_2 laser.

For the error calculation it is important to remark that increasing $\int_{magnet} Bdl$ (respectively the bending angle Θ), or increasing the distance L is equivalent. Since the choice of Θ is limited

---

[6]The center-of-gravity of the backscattered photons resembles precisely the position of the original beam at the crossing point.
[7]Eq. (6.7) is equivalent to Eq. (2.19), where $\int_{magnet} Bdl$ is replaced by the product Bl, with l the length of the magnet, $(L + l/2) \simeq L$ and the energy expressed in GeV.

6.3 Overview of the Energy Spectrometer

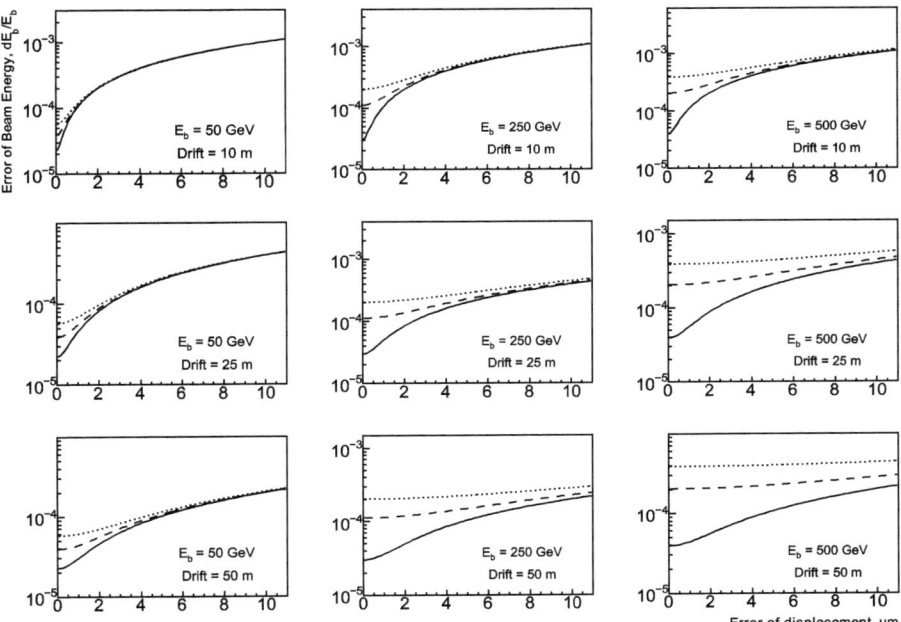

Figure 6.8: Beam energy uncertainty as a function of the error of the edge electron displacement for the green (dotted curve), infrared (dashed curve) and CO_2 laser (full curve) for three beam energies and three drift spaces.

by the beam emittance growth, it is appropriate to fix its value and vary L. Thus, assuming a relative error of the field integral of $2 \cdot 10^{-5}$ and for $\Delta L/L = 5 \cdot 10^{-6}$, Fig. 6.8 displays $\Delta E_b/E_b$ as a function of the error of the displacement ΔD for the three beam energies, three laser wavelengths and three values of L. The B-field integral is scaled such that the bending angle for the beam particles is 1 mrad at all energies.

From Fig. 6.8 some basic conclusions can be drawn. As already noted, a laser with larger wavelength provides a smaller error compared to lasers with smaller wavelength. Increasing the drift distance reduces $\Delta E_b/E_b$, while higher beam energies increase the error. The accuracy needed for the displacement should not exceed few micrometers for the CO_2 laser if L=50 m, for all E_b values. For $\Delta D \sim 1 \div 3 \mu$m, the error of the displacement is the dominant contribution to the beam energy uncertainty, thus a better knowledge of $\Delta B/B$ of e.g. $1 \cdot 10^{-5}$ provides only minor improvements at all energies.

In the present BDS [58], free drift space allows for lever arms of about 25 m. For such a drift distance, infrared and green lasers provide at 250 and 500 GeV $\Delta E_b/E_b$ values greater than 10^{-4} even for a perfect displacement measurement ($\Delta D = 0$), i.e. for measurements with infinite statistics and no systematics errors. In this case the dominant contribution to the error of the beam energy comes from the uncertainty of the integrated B-field.

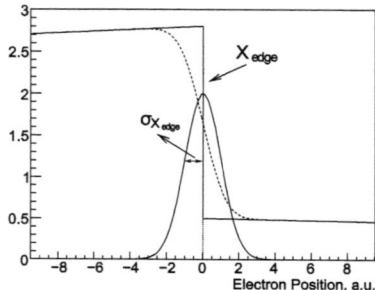

Figure 6.9: Sketch of the scattered electron position distribution near X_{edge}. The step function is convoluted with a Gaussian which contains all possible smearing effects. The resulting distribution is described by the dashed line.

Since the displacement is determined by the center-of-gravity of the recoil γ-rays and the position of the electron edge, the displacement error ΔD is given by the corresponding uncertainties as $\sqrt{\Delta X_\gamma^2 + \Delta X_{edge}^2}$, where ΔX_γ and ΔX_{edge} are the errors of X_{gamma} and X_{edge}, respectively, including statistical and systematic contributions.

Figure 6.9 illustrates the shape of the scattered electron position distribution near X_{edge}. Close to the edge, the ideal distribution is represented by a step function which has to be convoluted by a Gaussian to account for possible smearing effects due to, for example, the finite beam energy spread, beam size and detector resolution, shown by the dashed line. Hence, the statistical error of X_{edge} can be estimated as

$$\Delta X_{edge}^{stat} = \sqrt{\frac{2 \cdot \sigma_{X_{edge}}}{\frac{dN}{dx}(X_{edge})}} \,, \qquad (6.11)$$

where dN/dx is the scattered electron density at the detector plane and $\sigma_{X_{edge}}$ the width of the edge. After passing the spectrometer magnet the edge electrons are displaced from the beam electrons by an amount of $A \cdot 4\omega_0/m^2$, with $A \propto L \cdot \int Bdl$ and ω_0 as given in (6.5) (see also Eq. (6.15)), with a width very close to that of the beam at the detector plane. $\sigma_{X_{edge}}$ is uniquely determined by linac parameters such as beam size, energy spread, divergence, etc. With $A \propto L \cdot \int Bdl \propto L \cdot E_b$, we obtain

$$\sigma_{X_{edge}} \simeq \sqrt{\sigma_x^2 + \left(\frac{A}{E_b} \cdot \frac{\sigma_{E_b}}{E_b}\right)^2}$$
$$= \sqrt{\sigma_x^2 + \left(C \cdot L \cdot \frac{\sigma_{E_b}}{E_b}\right)^2} \,, \qquad (6.12)$$

with σ_x the horizontal bunch size at the detector plane and σ_{E_b}/E_b the relative energy spread of the beam and C a constant. Eq. (6.12) does not involve laser parameters because their contributions are much smaller or negligible.

6.3 Overview of the Energy Spectrometer

Using beam values as discussed in Sect. 6.4, $\sigma_{X_{edge}}$ is estimated to be in the range of $70 \div 90$ µm. In our approach, the edge distribution is assumed to be described by a convolution of a Gaussian with a step function, however any other ansatz can be used.

For 10^6 Compton scatters, ΔX_{edge} turns out to be in the order of 6 µm for an infrared laser, so that together with $\Delta X_\gamma = 1$ µm (Sect. 6.4), the displacement error is close to 7 µm, and somewhat larger for a green laser. Therefore, if the approach illustrated in this section is followed, the use of a CO_2 laser is favored and excludes (with high confidence) operation of lasers with smaller wavelengths.

A peculiar problem which we have to account for is the amount of synchrotron radiation generated when the beam electrons pass through the dipole magnet and its possible impact on precise position measurements. This will be discussed in Sect. 6.4.

6.3.3 Method B

Beam and Compton scattered electrons with energy E propagate to the detector such that their transverse position is well approximated by

$$X(E) = X_0 + \frac{A}{E}, \qquad (6.13)$$

where $A \sim L \cdot \int Bdl$ and X_0 the position of the original beam line extrapolated to the detector plane, which is given by the center-of-gravity of the backscattered γ-rays, X_γ. Note that in (6.13) small effects related to synchrotron radiation are omitted.

According to Eqs. (6.6) and (6.13), the positions of the beam and edge electrons can be expressed as

$$X_{beam} \equiv X(E_{beam}) = X_\gamma + A/E_{beam} \qquad (6.14)$$

$$X_{edge} \equiv X(E_{edge}) = X_{beam} + A \cdot \frac{4\omega_0}{m^2}. \qquad (6.15)$$

Hence, the beam energy can be deduced from

$$E_b = \frac{m^2}{4\omega_0} \cdot \frac{X_{edge} - X_{beam}}{X_{beam} - X_\gamma}. \qquad (6.16)$$

Thus, instead of recording the energy of the edge electrons, the beam energy can be accessed from measurements of three particle positions, the position of the forward going backscattered γ-rays, the position of the edge electrons and the position of the beam particles. The position X_{beam} can be measured by a beam position monitor (BPM), while recording X_{edge} and X_γ needs dedicated high spatial resolution detectors very similar to the demands of method A. Besides the limitation to a CO_2 laser for the concept of edge energy measurements (method A), the demand of $2 \cdot 10^{-5}$ for the field integral uncertainty is rather challenging, and less stringent requirements would be of great advantage. In method B, E_b determination does not depend on the field integral, the length of the magnet as well as the distance to the detector plane. In particular, the independence on the integrated B-field only requires a rather a coarse $\Delta B/B$ monitoring. It is, however, necessary to ensure that both the beam and the edge electrons have to pass through the same B-field integral, i.e. the magnetic field has to be uniform across the large bending range. Also, the distance $X_{edge} - X_{beam}$ in (6.16) which involves as a product the integrated B-field and the sum of the drift distance and the length of the magnet [59], does not depend on the beam energy. Possible variations of this distance may only be caused by rather slow processes of environmental nature. Thereby, by accumulation of many bunch related $X_{edge} - X_{beam}$ measurements, high statistical precision can be achieved for this quantity. This implies the option to operate the spectrometer with lasers of less pulse power, which is of great advantage since the laser pulse power is a critical issue for method A. The novel approach of recording three particle positions (the three-point concept) seems therefore to be a very promising alternative[8].

Also, Eq. (6.16) reveals that due to the proportionality between the beam energy and the distance $X_{edge} - X_{beam}$, which is larger as smaller the wavelength of the laser, best beam energy values are obtained for high energy lasers, a situation which is opposite to that of method A.

The precision of the beam energy can be estimated as

$$\frac{\Delta E_b}{E_b} = \frac{X_{edge}}{X_{edge} - X_{beam}} \left(\frac{\Delta X_{edge}}{X_{edge}} \right) \oplus$$
$$\oplus \frac{X_{edge}}{X_{edge} - X_{beam}} \left(\frac{\Delta X_{beam}}{X_{beam}} \right) \oplus \frac{\Delta X_\gamma}{X_{beam}} . \qquad (6.17)$$

Here the three terms have to be added in quadrature. Assuming for the crossing angle 8 mrad and (achievable) values for $\Delta X_{beam} = 1$ μm and $\Delta X_\gamma = 1$ μm, expected beam energy uncertainties are shown against the edge position error, ΔX_{edge}, for the CO_2, infrared and green lasers in Fig. 6.10, in analogy to Fig. 6.8. Drift distances of 10, 25 and 50 m and beam energies of 50, 250 and 500 GeV are supposed. The error on the beam energy increases for decreasing drift distances and beam energies. Furthermore, a laser with smaller wavelength is favored, since a CO_2 laser exceeds the value of 10^{-4} for $\Delta E_b/E_b$ very quickly. Considering in particular the case of 250 GeV beam energy and 25 m drift distance, one notices that even for a perfect X_{edge} determination ($\Delta X_{edge} = 0$), the beam energy uncertainty for a CO_2 laser is larger than 10^{-4}.

Table 6.1 collects the contributions of the three terms of Eq. (6.17) for two drift distances and three beam energies. The laser used is a green laser and $\Delta X_{beam} = \Delta X_\gamma = 1$ μm is assumed. Since the statistical fraction of ΔX_{edge} is varying as functions of energy and drift distance as well (see Eq. (6.11)), different values for ΔX_{edge} were accounted for. These values are given by the geometrical sum of the statistical error from Eq. (6.11) and a systematic error of 4 μm

---
[8] Also, vice versa, knowing $X_{edge} - X_{beam}$ with high precision, the B-field integral can be deduced with similar accuracy.

6.3 Overview of the Energy Spectrometer

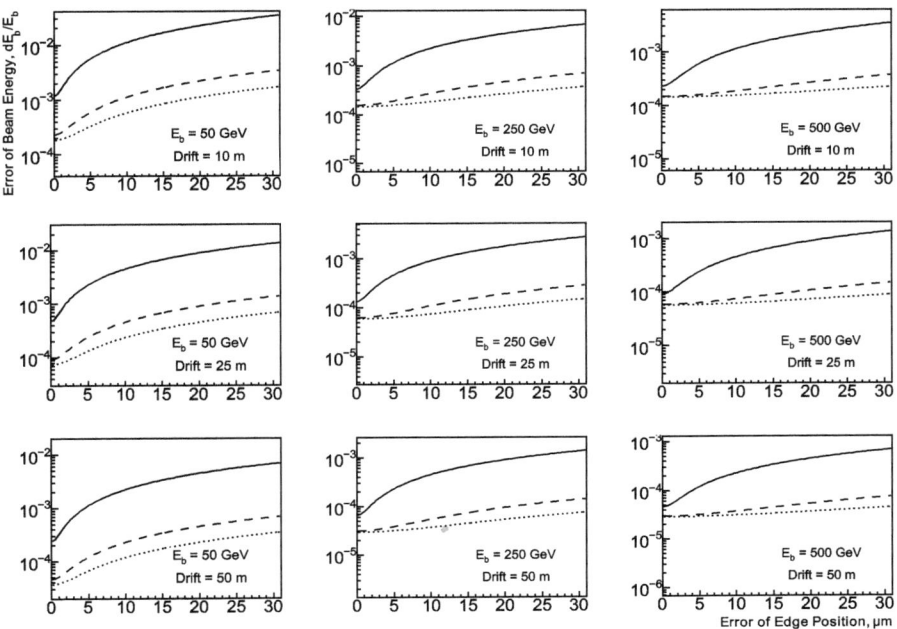

Figure 6.10: Beam energy uncertainty as a function of the edge position error for the green (dotted curve), the infrared (dashed curve) and the CO_2 laser (full curve) for three beam energies and drift spaces.

Beam Energy (GeV)		50		250		500	
Distance L (m)		25	50	25	50	25	50
ΔX_{edge} (μm)		5	6	10	15	15	24
$\frac{X_{edge}}{X_{edge}-X_{beam}}\left(\frac{\Delta X_{edge}}{X_{edge}}\right)$	(ppm)	111	66	44	33	33	26
$\frac{X_{edge}}{X_{edge}-X_{beam}}\left(\frac{\Delta X_{beam}}{X_{beam}}\right)$	(ppm)	62	31	44	22	42	20
$\frac{\Delta X_0}{X_{beam}}$	(ppm)	40	20	40	20	40	20

Table 6.1: The individual contributions (in parts per million) of the three terms in Eq. (6.17) for a green laser, for two drift distances, three beam energies and different values of ΔX_{edge}. Values for $\Delta X_{beam} = 1$ μm and $\Delta X_\gamma = 1$ μm are assumed.

which is assumed to be independent on the drift distance and beam energy.

As can be seen, the beam energy accuracy for 25 m and 50 GeV results in 133 ppm, which somewhat exceeds the limit of 100 ppm. Improvements are possible by choosing a larger distance or, equivalently, a higher B-field integral or more statistics. At higher beam energies, the energy uncertainties are well below the anticipated limit.

6.4 Detector Options and Simulation Studies

The detector assembly is assumed to be located at least 25 m downstream of the magnet. Since we plan to operate the spectrometer with an energy independent fixed bending angle of 1 mrad, the distance of the backscattered γ-ray centroid to the beam line 25 m downstream of the dipole is 26 mm for all E_b values, while the displacement of the edge electrons depends on E_b and E_λ. This displacement in the range of a few centimeters to about a quarter of a meter requires high stability of the detector assembly and its adjustment to micrometer accuracy. Therefore, the individual detector components should be connected rigidly and installed on a vibration damped table that can be moved horizontally (and vertically) and controlled with high precision.

After leaving the vacuum chamber, the Compton scattered electrons near the edge traverse a position sensitive detector with high spatial resolution. We propose to employ either a diamond micro-strip or an optical quartz fiber detector. Such detectors, frequently applied in particle physics experiments, have demonstrated their ability to achieve $1 \div 10$ μm spatial resolution within an intense radiation field, see e.g. [60, 61].

6.4.1 Photon Detection

The center-of-gravity of the Compton γ-rays might be recorded by employing one of the two following concepts. One concept consists in measuring high energy electrons and positrons from photon interactions in a converter placed closely in front of the tracking device. According to simulations, a tungsten converter[9] of sufficient radiation lengths seems to be suitable. Compton backscattered photons will be affected inside the converter by several processes, creating an electromagnetic shower. The particles leaving the converter (electrons, positrons and photons) have a spatial distribution with a center-of-gravity which is expected to coincide with that of the original backscattered photon burst. Such a scheme constitutes some trade-off between large conversion rates and accurate photon position determinations, which might be altered by multiple scattering of the forward collimated $e^\pm$ particles within the converter. As a position sensitive detector a quartz fiber detector similar to that for edge position measurements is proposed and, as simulation studies revealed, sub-micrometer precisions of the original photon position are achievable. In a quartz fiber detector, the signal is generated by Cerenkov photons emitted by charged particles with energy above some threshold value.

Together with backscattered Compton photons, synchrotron radiation will be generated by electrons passing through the magnet. For the magnet as described in [13], about five photons per beam particle with an average energy of 3.8 MeV are generated with an energy spectrum that peaks below 1 MeV, resulting in a total number of 10^{11} γ's per bunch. They are concentrated within the cone of the forward produced Compton scattered photons and the bent beam. If

---

[9]Tungsten with its large atomic number of 74 and high density of 19.3 g/cm^3 is an attractive material for small converters. However, pure material is difficult to cast or machine, but powder metallurgy processes can produce a sintered form of tungsten, with a density only slightly below that of the pure metal.

6.4 Detector Options and Simulation Studies

a tungsten converter of e.g. 16 radiation lengths (X_0) in front of the X_γ detector is inserted, it also serves as an effective shield against SR. However, the huge amount of such photons (plus a minor fraction from Compton scattered electrons) may preclude perfect SR protection. Possible low energy electrons and positrons from SR showers are expected to enter the detector and could modify the response and eventually the center-of-gravity of the primary Compton photons. The impact of this background (together with machine related background) has to be taken into account in procedures of precise X_γ determinations. Properties of particles leaving the converter and prescriptions addressed to eliminate center-of-gravity distortions are discussed in Sect. 6.4.2.

The converter is assumed to have a cross-section of 2×2 cm^2 and a length of 16 X_0. The transverse dimension of the converter is mainly dictated by the small displacement of the beam particles 25 m downstream of the spectrometer magnet. A converter of e.g. 26 X_0 with more efficient SR removal results in some less precise γ-centroid measurements and is considered to be less favored.

A completely different way to record the undeflected beam position relies on monitoring the edge of SR light at X = 0, without a converter in front of the position device. Dedicated and novel SR devices were suggested in [51]. In this thesis, it is proposed to employ an avalanche detector with gas amplification. SR light which passes a 10×10 mm^2 entrance window of 1 mm beryllium[10] generates an avalanche in xenon gas at 60 atm over a range of 1.5 mm, the gap between the anode and cathode. The transverse size of the avalanche is expected to be close or below 1 μm, and due to the amplification process, a large number of electrons is produced and generates a sufficiently strong output signal [51]. The anode plane of the detector consists of 1 μm nickel layers with 2 μm dielectric separation in between. Such a geometry matches very well the transverse size of the avalanche and permits sub-micrometer access of the position of the SR edge. Since no converter is needed in this scheme, the 10^6 high energy Compton photons are now background. Their impact on the accuracy of the SR edge is negligible as will be shown below.

Discussions on whether recording the incident beam position by means of SR is superior to the conventional converter approach are also included in this section.

6.4.2 Simulation Studies

A full Monte Carlo simulation based on the GEANT toolkit [62] [11] has been developed to analyze the basic properties of the Compton spectrometer and to evaluate design parameters for the detectors. Bunches of $2 \cdot 10^{10}$ electrons are colliding with unpolarized or circular polarized infrared or green laser pulses of 10 ps duration by a generator inside the program[12]. The

---

[10] The beryllium foil also acts as the high-voltage cathode plane.

[11] At the beginning of the study GEANT3 (version 3.21/14) has been used, while later on GEANT4 (version 4.8.2) was implemented.

[12] Operating with a CO_2 laser requires larger free drift space than available in the present BDS. Therefore, no simulation results are presented for this case.

generator accounts for an internal electron bunch energy spread of 0.15% which is slightly larger than the values given in [9][13], a transverse bunch profile of 50 μm and 5 μm in horizontal, respectively, vertical direction and a 300 μm extension along the beam direction, all of Gaussian shape. Such input parameters are in accord with ILC beam properties within the BDS. A high-power pulsed laser with either $E_\lambda = 1.165$ eV or 2.33 eV is focused onto the incident beam with a crossing angle of 8 mrad. Also, perfect laser pointing stability and instantaneous laser power are assumed. As default event rate, 10^6 Compton scatters are generated for single bunch crossing.

Compton recoil electrons and photons as well as non-interacting beam particles are tracked through the spectrometer and recorded by the detectors. The magnet provides a fixed bend of 1 mrad for all beam energies anticipated. At the nominal energy of 250 GeV, the B-field integral corresponds to 0.84 T·m for a magnet length of 3 m. Synchrotron radiation is also taken into account. The position sensitive detectors which perform X_γ, X_{beam} and X_{edge} measurements are located 25 m downstream of the spectrometer magnet.

6.4.2.1 X_{edge} Determination

For the edge electrons, we assume either a diamond strip or a quartz fiber detector[14]. Both detector options have a transverse size of 1×1 cm^2. For the 300 μm thick diamond detector a pitch of 25 μm and a strip width of 10 μm were chosen. Possible crosstalk was also considered. When passing through a thin layer of matter, charged particles lose energy according to a distribution with a long tail at high energies (similar to a Landau distribution). Thereby, in rare cases the electron transfers a large amount of energy within the sensor which implies a large charge signal. A code based on GEANT has been written that simulates all physical processes taking place in the diamond strip detector (DSD) and calculates the energy deposited along the particle track in the detector[15]. The resulting deposited energy is used to weight each electron and, after summation over all entries in a given channel, the total signal is shown in the corresponding Fig.6.11 for the green (left) and infrared (right) laser.

For the quartz fiber detector (QFD), Compton electrons are measured by a single layer of 50 μm square fibers. A cladding thickness of 5 μm on each side results in an active fiber core of 40 μm. Crosstalk was considered to be negligible. Since only a fraction of typically a few percent of the light produced in the fibers is trapped and transported to the light detector, the small probability to detect a minimum ionizing particle is to great extent compensated by the large number of electrons traversing a single fiber. Therefore, despite a small single-particle

---

[13] The ILC Reference Design Report lists for the relative energy spread 0.14 and 0.10% for the electrons, respectively, positrons. The larger value for the electrons is due to their passage through a long undulator.

[14] Due to the expected large radiation dose, a silicon strip detector will not be considered here unless very radiation hard Si detectors become available.

[15] In general, the charge signal depends on the energy deposited along the track rather than the energy loss. Some of the energy lost by the particle is carried away by secondary electrons or by Cerenkov radiation.

6.4 Detector Options and Simulation Studies

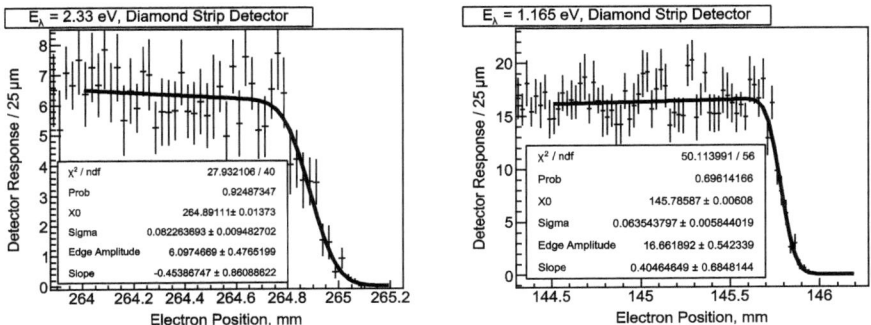

Figure 6.11: Diamond strip detector response for the green (left) and infrared laser (right).

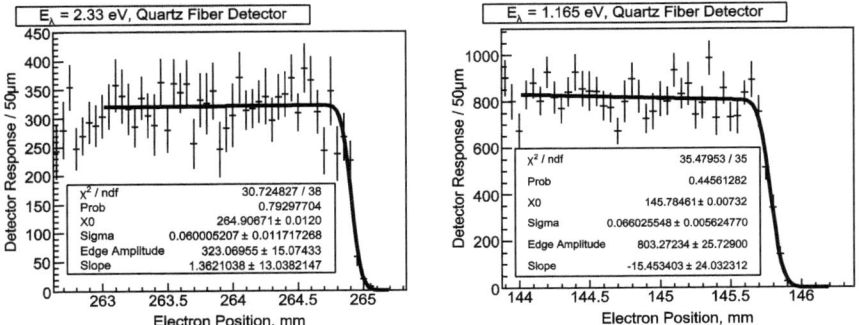

Figure 6.12: Quartz fiber detector response for the green (left) and infrared laser (right).

light yield, a detection efficiency for individual fibers of 100% was assumed.

The quartz fiber response was simulated by counting the number of Cerenkov photons generated by each electron along its path through the detector. The sum over all such photons within a fiber is proportional to the output signal and is plotted in the Fig. 6.12 for the green (left) and infrared (right) laser.

As can be seen, the expected sharp edges of the spectra are somewhat diluted, mainly due to the energy spread of the beam particles and the beam spot size. The edge positions of the spectra were obtained by a fit of an ansatz which results from a step-function plus a (uniform) background folded by a Gaussian as proposed in e.g. [28, 29, 31]:

$$G(x, p_1, p_2, ..., p_6) = \frac{1}{2}(p_3 + p_4(x - p_1)) \cdot \text{erfc}\left[\frac{x - p_1}{\sqrt{2}p_2}\right]$$
$$- \frac{p_2 p_4}{\sqrt{2\pi}} \cdot \exp\left[\frac{(x - p_1)^2}{2p_2^2}\right] + p_5 + p_6(x - p_1) \,. \quad (6.18)$$

The edge position p_1, the edge width p_2, the amplitude of the edge p_3, the slope p_4, the background level p_5 and its slope p_6 were treated as free parameters. Assuming $p_5 = p_6 = 0$ in our particular case, the errors of the edge positions were found in the range of 5 to 15 μm, with values of 6 and 7 (12 and 14) μm for the infrared (green) laser. These numbers are in accord with the endpoint position demands shown in e.g. Fig. 6.10 and Tab. 6.1 for the approach of recording three particle positions, X_γ, X_{beam} and X_{edge}. It is also evident that method A based on direct edge energy measurements (by means of precise B-field integral and edge displacement information) seems to be not-favored: precisions of edge electron displacements of a fraction of a micrometer up to only few micrometers (see Fig. 6.8) are difficult to achieve without additional effort.

In principle, beam polarization may affect the endpoint p_1 which might be coupled with the slope of the energy spectrum p_4 in the vicinity of the edge position as indicated in Fig. 6.2. By Compton simulation of 80% polarized electrons of 250 GeV with circular polarized infrared laser light we found that the edge position differs by less than 1 μm with respect to the case of unpolarized electrons. Thereby, Compton scattering of polarized beams will not noticeably modify p_1 and hence the beam energy measurement.

The assumption of a Gaussian internal energy spread rest on ongoing machine design studies. As long as collective effects as intra beam scattering (IBS) or interactions with the vacuum chamber impedance are negligible, the energy spread is expected to be of Gaussian shape. Since at present a final design of the vacuum chamber to minimize the beam impedance and IBS effects is not completed a realistic shape of the energy distribution is missing. Deviations from a Gaussian, if any, are however expected to be small [63]. Preliminary accelerator simulations reveal that the energy spread is close to a Gaussian distribution [64] and support our assumption. This holds for the electrons as well as the positrons despite different sizes of the relative energy spread. If it will be demonstrated by measurements that the energy spread is not Gaussian distributed, the fitting function (6.18) has to be modified accordingly.

As estimated in [13], the radiation dose for both detectors is well below the tolerance limits.

6.4.2.2 X_γ Determination

As already introduced in Sect. 6.4.1, within one approach of measuring X_γ, the center-of-gravity of the Compton scattered γ-rays is inferred indirectly via conversion to electrons and positrons within a 16 radiation lengths tungsten converter. When entering the converter, the photons are concentrated within a spot of approximately 250 μm r.m.s., a size which is dominated by the $\sim 1/\gamma$ angular distribution of the Compton process. After a first estimate of the thickness of the converter, a full simulation of the 56 mm long conversion material has been performed. In particular, the process of converting the 10^6 Compton photons together with the SR photons along with the trajectories of the resulting electrons and positrons through the converter and into the fiber detector was simulated. Despite the small transverse extension of the converter, the core of the shower particles caused by Compton photons is assumed to maintain the initial γ-centroid position (being at X = 0 in the simulation). Directly after the converter the quartz fiber detector array of 50 μm fibers has been placed in order to measure the $e^\pm$ shower particles

6.4 Detector Options and Simulation Studies

from which the γ-centroid position has to be deduced. Figure 6.13 (left) shows the number of charged particles escaping the converter as a function of X, while their energy behavior is shown on the right-hand side[16]. The spectra indicated as 'Signal' are $e^{\pm}$ particles from Compton photons, whereas those marked as 'Background' are from synchrotron radiation. We expect $1.5 \cdot 10^8$ charged particles from 10^6 Compton events, with an average energy of 25.8 MeV. Their density distribution, dN/dx, clearly peaks at X = 0.

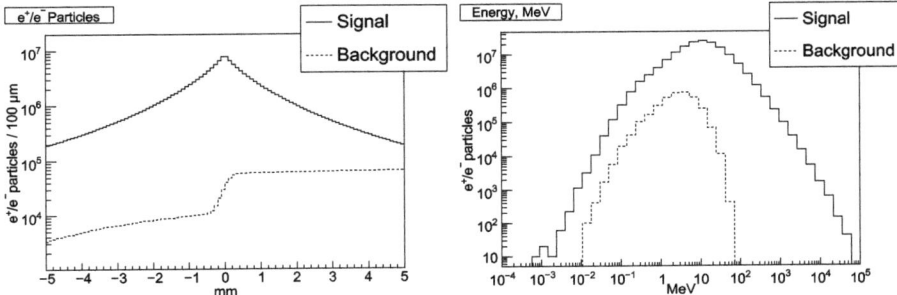

Figure 6.13: Left: Number of charged particles escaping the 16 radiation lengths tungsten converter as a function of X. Right: Energy distribution of charged particles escaping the converter. The 'signal' spectra are normalized to 10^6 Compton scatters, while the 'background' spectra are normalized to $2 \cdot 10^{10}$ beam particles within a bunch.

Besides charged particles, photons also escape the converter. They are either generated within electromagnetic showers from Compton scattered and SR γ-rays or are SR photons which pass the converter without interaction. A fraction of less than 2% of the original SR yield with an average energy of 3.9 MeV survives. Their dN/dx and energy spectra are shown in Fig. 6.14.

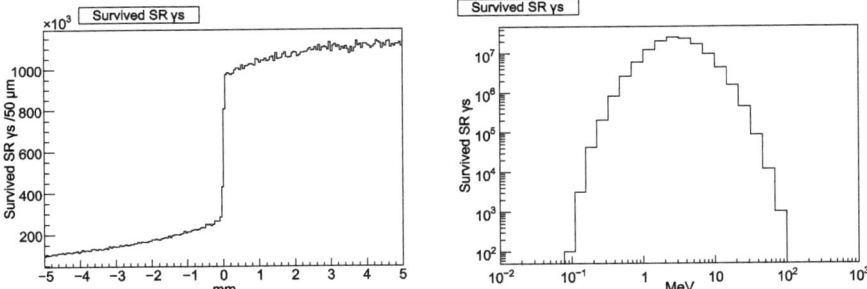

Figure 6.14: Left: Number of SR γ-rays which escape the 16 radiation lengths tungsten converter as a function of X. Right: Energy distribution of SR γ-rays escaping the converter. Both spectra are normalized to $2 \cdot 10^{10}$ beam particles within a bunch.

For the position sensitive X_γ device, a single layer of quartz fibers is assumed with properties identical to those for the edge electron detector. Basically, this detector should have a large

---
[16] Analogous spectra are obtained for the vertical direction as well as if the infrared laser is replaced by the green laser.

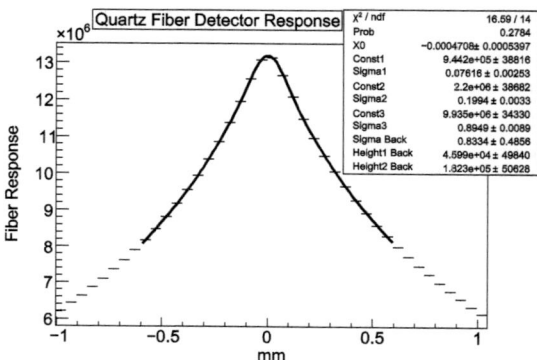

Figure 6.15: Cerenkov light response of all charged particles passing the quartz fiber detector. The curve is the result of a fit of the sum of three Gaussian distributions and the step function in Eq. (6.18) with $p_4 = p_6 = 0$.

sensitivity to charged particles from pair production of Compton photons within the converter and 'blind' with respect to background (SR) γ-rays. In Fig. 6.15 the response function of the detector in terms of the amount of Cerenkov light generated from all $e^{\pm}$ particles within a fiber is shown together with the result of a fit. The fit result is based on a two-step procedure. First, due to a priori unknown precise γ-centroid position, X_{γ} is approximately determined by a simple algorithm [65], which fixes the peak position within about ± 25 μm. Then, selecting a fitting range of some ± 600 μm around this preliminary centroid, an empirical fit of the sum of three Gaussians and the step function in Eq. (6.18), with $p_4 = p_6 = 0$, provides the ultimate peak position of $X_{\gamma} = $ -0.47 ± 0.54 μm with a $\chi^2/NDF = 16.59/14$, corresponding to 27.8% probability[17].

The fit range chosen excludes particles which are less sensitive to the peak position but sensitive to the background. The peak value found is in good agreement with the expectation of zero and its error is less than the anticipated limit of ~ 1 μm. The distribution in Fig. 6.15 is the response of all escaping $e^{\pm}$ particles generated from 10^6 Compton photons and the appropriate fraction of SR, after normalization to $2 \cdot 10^{10}$ beam electrons. The latter causes a slight asymmetry with respect to X = 0 and is the reason to include the step function within the fit. As a consequence, a rather complicated response behavior is obtained and after some trials the spectrum was reasonably described by the selected ansatz. If instead of a 50 μm fine segmented detector an array of 100 μm quartz fibers is utilized the centroid position and its error are found to be in agreement with the values quoted above. Irrespectively of the details for the final design of the converter-fiber detector system, this option seems to be capable to meet the requirements, in particular if instead of only one fiber layer several layers with some staggering are employed.

A different approach to record the incident beam direction consists of using a SR edge

---

[17] If the fit is performed with the sum of only two Gaussians and the step function, the χ^2/NDF is significantly worse.

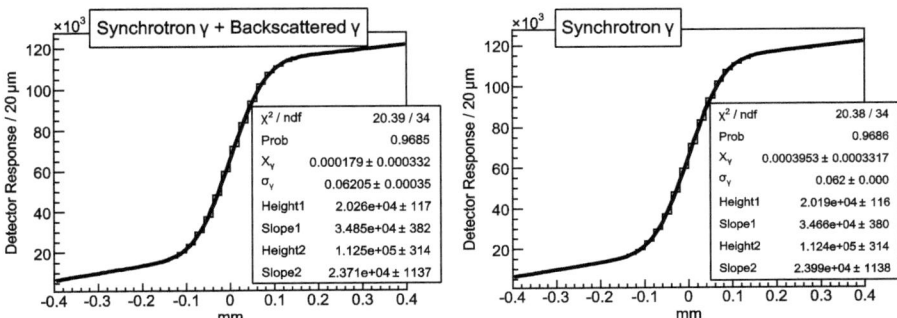

Figure 6.16: Left: Response function of the avalanche SR edge detector for all signal and background photons. Right: Response function of the same detector for only SR signal photons. The curves are the results of a fit of Eq. (6.18) supplemented by an additional background tail.

detector. The avalanche detector of Ref. [51] with xenon being in a superfluid state with a density of 3.05 g/cm^3 is proposed to perform SR edge position measurements around X = 0. A detector acceptance of ±5 mm will be exposed by some 20% of the 10^{11} SR photons and all 10^6 Compton recoil γ-rays, which are considered now as background. Photons traversing the detector interact with the xenon so that electrons are created via e.g. the photoelectric effect or pair production. These electrons drift toward the anode and in collisions with xenon atoms they liberate further electrons. This process is accompanied by loss of energy of the electrons and deflection from their incident direction. The response of such a detector was simulated and the X-position of each electron-atom collision weighted by the corresponding released energy is plotted in Fig. 6.16 for all photons (left) and only the SR γ-rays (right). Clearly, the SR edge at X = 0 is well recognized and a fit using Eq. (6.18) provides $X_\gamma = 0.18 \pm 0.33 \mu$m. This number is, despite of the crudeness of the simulation, in perfect agreement with the demands and indicates that the response of Compton photons as background is not important. Hence, X_γ position measurements can be performed with an avalanche SR detector as proposed in [51]. Presently, such detector does not exist, but R&D is ongoing and first results are expected in 2009/10 [66].

6.5 Laser Power

So far we assumed 10^6 Compton interactions per crossing regardless of the laser type used. To achieve such an event rate, the required laser power is estimated as follows. Assuming for the incident electron beam transverse bunch sizes of $\sigma_x = 20$ μm, $\sigma_y = 2$ μm and 300 μm in longitudinal direction at the Compton IP, for the transverse laser spot size 100 (50) μm in the case of an infrared (green) laser, a pulse duration of 10 ps, a crossing angle of 8 mrad and

$2 \cdot 10^{10}$ electrons per bunch, the infrared laser $e\gamma$ luminosity per crossing is according to [67]

$$L_{pul} = N_\gamma \cdot N_e \cdot g \,, \qquad (6.19)$$

with N_γ the number of photons per laser pulse and N_e the number of electrons per bunch. The geometrical factor g for vertical beam crossing[18] is well approximated by

$$g = \frac{\cos^2 \alpha/2}{2\pi} \cdot \frac{1}{\sqrt{\sigma_{xe}^2 + \sigma_{x\gamma}^2}} \cdot \frac{1}{\sqrt{(\sigma_{ye}^2 + \sigma_{y\gamma}^2)\cos^2(\alpha/2) + (\sigma_{ze}^2 + \sigma_{z\gamma}^2)\sin^2(\alpha/2)}} \,, \qquad (6.20)$$

where α is the crossing angle and the transverse laser profile is assumed to be constant. Note that the vertical and longitudinal bunch sizes $\sigma_{y\gamma}$, σ_{ye} and $\sigma_{z\gamma}$, σ_{ze}, respectively, of the interacting beams contribute. Thus, the luminosity per crossing results in 0.166 per mbarn and μJ, while a green laser provides 0.307 mb$^{-1}\mu$J^{-1} [19]. If these luminosities are combined with the corresponding Compton cross section of $\sigma = 197.9$ mb, respectively, 137.7 mb, a bunch related laser power of 30 or 24 mJ is obtained. At present, such lasers that match the pattern of the incident electron bunches are not commercially available. But the FLASH collaboration [56, 68], employing a laser in the infrared region with good reliability, and ongoing R&D for green lasers within the ILC community [57] will set milestones in the future, from which these studies could greatly benefit.

6.6 Potential Background Processes

Usually, the characteristics of Compton scattering are calculated within the Born approximation, see Sect. 6.2 as an example. Compton scattering at the ILC with large bunch densities, large laser flash energies and small pulse lengths ensures sufficient $e\gamma$ luminosity, which is important for precise E_b determination. However, under such conditions, multiple scattering, nonlinear QED effects, the Breit-Wheeler process (also called two-step pair production) and higher order QED corrections can contribute and might have a non-negligible impact on E_b measurements. In this section, these effects will be discussed and their influence on energy measurements evaluated. In the following, if not explicitly stated, the input parameters in Tab. 6.2 are assumed

The laser power is supposed to ensure 10^6 Compton events at any energy, e.g. at 250 GeV it is 0.4 mJ for a CO_2 and 30 mJ for a green laser.

6.6.1 Multiple Scattering

When the thickness of the laser target is about one collision length, each electron may undergo multiple Compton scattering within the crossing region. The probability might not be small

---
[18] For horizontal crossing, the roles of x and y have to be interchanged.
[19] Shortening the pulse duration to 5 ps increases the luminosity by only 0.6% (2.3%) for infrared (green) laser operation.

6.6 Potential Background Processes

Laser/beam parameters	Value	Unit
$\sigma_{x\gamma}, \sigma_{y\gamma}, \sigma_{z\gamma}$	50, 50, 10	μm, μm, ps
$\sigma_{xe}, \sigma_{ye}, \sigma_{ze}$	20, 2, 300	μm, μm, μm
Crossing angle α	8	mrad
Magnet length	3	m
Magnet field	0.28	T
Free drift distance	25	m
Compton events	10^6	electrons/photons

Table 6.2: Input laser/beam parameters as assumed in the simulation.

because, after a large energy loss in a first collision, the Compton cross-section increases and together with the high particle densities of the colliding bunches further collisions can be caused. Such multiple scattering leads to a low energy tail in the energy spectrum of the scattered electrons and could modify the sharp edge behavior.

In a first step, the fraction of incident electrons which scatter several time with laser photons is evaluated and, in a second step, we evaluate the energy spectrum of such electrons and their impact on the endpoint position.

6.6.1.1 Rate of Electrons with Multiple Interactions

In order to understand the procedure used to evaluate the number of electrons with multiple interactions a short introduction is appropriate.

When N_e electrons hit a fixed target of thickness δx, the number of scattered electrons N'_e is

$$N'_e = N_e(n\sigma\delta x) , \qquad (6.21)$$

where n is the density of the scattering centers and σ the cross-section. The formula is valid for $N'_e \ll N_e$. Note that N'_e is proportional to the density of the scattering centers.

The probability for an electron to suffer a collision between x and $x + dx$ is

$$F(x) = (1 - n\sigma x)w dx , \qquad (6.22)$$

with $w = n\sigma = 1/\lambda_{path}$ and λ_{path} the mean free path. For $N'_e \ll N_e$, the target thickness is much smaller than the mean free path, $\delta x \ll \lambda_{path}$, and the mean position where the electrons scatter is given as

$$\bar{x} = \frac{\int_0^{\delta x} xF(x)dx}{\int_0^{\delta x} F(x)dx} \sim \frac{\delta x}{2} , \qquad (6.23)$$

which is independent from the density of the scattering centers..

After the first collision the electron continues to travel across the target and can scatter at least a second time. The number of such electrons is

$$N''_e = N'_e n \bar{\sigma} \frac{\delta x}{2}, \qquad (6.24)$$

since $N'_e \propto n$, N''_e is proportional to n^2. The electrons scattered once have a non-monochromatic energy spectrum, so that $\bar{\sigma}$ is the cross-section weighted by the energy spectrum of the electrons.

In our case, the target is the laser pulse and the density of the scattering centers is not a constant but of Gaussian shape, which, however, does not change the basic results.

It is in general difficult to determine directly the number of electrons which scatter at least once inside the laser bunch (N'_e) and those which scatter at least twice (N''_e) by means of the program CAIN [69], because the laser power assumed is too low. Therefore, these numbers were determined by an extrapolation procedure. For that it is sufficient to evaluate N'_e and N''_e for a laser with higher photon density as assumed so far (or, equivalently, with more laser power if all other parameters are unaltered). We expect, according to Eqs. (6.21) and (6.24), a linear, respectively, quadratic behavior for N'_e and N''_e on the laser power.

For CO_2 laser powers between 0.01 J and 0.3 J, the number of electrons with generation[20] ≥ 2 (N'_e) and ≥ 3 (N''_e) were counted and Fig. 6.17 shows the results.

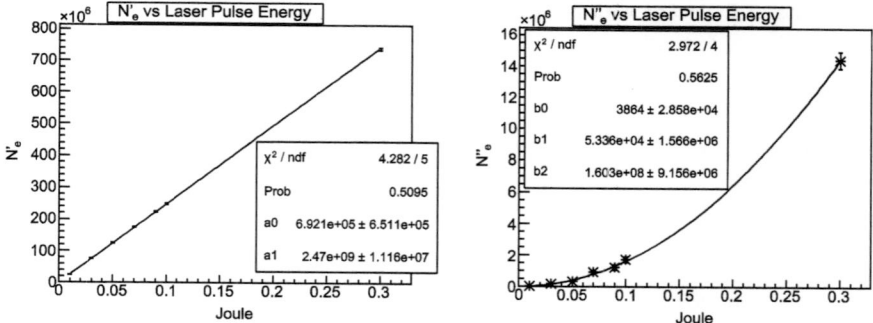

Figure 6.17: Left: Number of electrons which scatter at least once inside the laser bunch (generation ≥ 2) vs. laser power. Right: Number of the electrons which scatter at least twice (generation ≥ 3) vs. laser power (right).

The numbers on Fig. 6.17 (left) were fitted with a straight line ($y = a_1 x + a_0$), while those on Fig. 6.17 (right) with a parabola ($y = b_2 x^2 + b_1 x + b_0$), according to the hypothesis of a linear, respectively, quadratic dependence of the number of scattered electrons on the laser power.

From the slope of the linear fit and the coefficient b_2 of the parabola (all other parameters

---

[20] A generation is defined as the number of interactions suffered by a particle incremented by one, e.g. the non-interacting beam particles have generation of 1

6.6 Potential Background Processes

are negligible) we obtain

$$N'_e = a_1 \cdot P_{laser}$$
$$= (2.47 \cdot 10^9) \cdot (4 \cdot 10^{-4}) \simeq 10^6 \qquad (6.25)$$
$$N''_e = b_2 \cdot P_{laser}^2$$
$$= (1.6 \cdot 10^8) \cdot (4 \cdot 10^{-4})^2 \simeq 26 \;, \qquad (6.26)$$

and finally

$$\frac{N''_e}{N'_e} \simeq 26 \cdot 10^{-6} \;. \qquad (6.27)$$

The number of electrons which scatter at least three times were also evaluated and found to be negligible. For a CO_2 laser with a huge pulse power of 0.1 J, as an example, $N'''_e/N'_e < 8 \cdot 10^{-5}$.

6.6.1.2 Spectra of electrons with multiple scattering

The energy spectrum of electrons which scatter at least twice was extracted from events generated by CAIN assuming a laser power of 0.3 J. In good approximation, the spectrum is independent on the laser power. This spectrum was saved in a root file, while for electrons which scattered only once the standard GEANT generator was employed. The energy and position of both types of electrons at the detector plane are shown in Fig. 6.18 for a CO_2 laser. Close to $X_{E_{edge}}$, only few electrons which scattered twice are expected. We conclude that multiple scattering effects are very small and any impact on the edge position is practically not measurable. Additional simulations were performed for a green laser. The transverse and longitudinal laser and beam profiles were identical to those for the CO_2 laser. For 50 and 250 GeV energies and a laser power of 11 mJ, respectively, 30 mJ, N''_e of $\sim$30 for Z-pole running and $\sim$70 for the nominal beam energy were found. The laser power was adjusted such that 10^6 Compton scatters are generated in both cases.

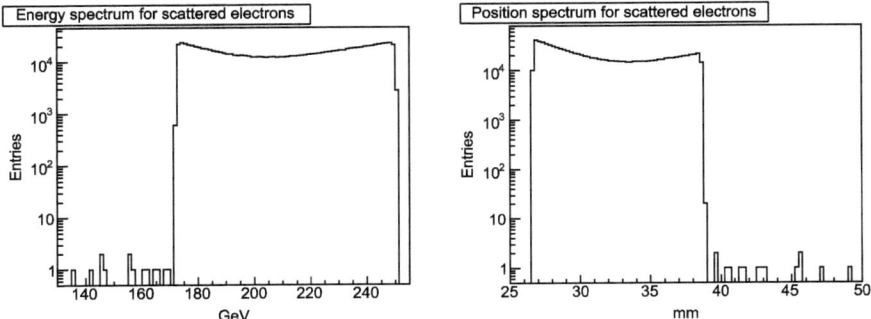

Figure 6.18: Left: Energy spectrum including electrons which scattered at least twice. Right: Position distribution of such electrons at the detector.

So far unpolarized electrons beam were assumed, since polarization plays a minor role in this estimation. In fact, the total cross-section for Compton scattering can be written as

$$\sigma_c = \sigma_c^0 + \lambda P_e \sigma_c^1 , \tag{6.28}$$

with σ_c^0 the unpolarized and σ_c^1 the polarized cross-section, P_e and λ the mean helicity of electron beam and laser light, respectively. Figure 6.19 shows for three laser energies the ratio σ_c^1/σ_c^0 as a function of E_b. As can be seen, σ_c^1/σ_c^0 $mathrel\lesssim 0.25$ for all cases considered.

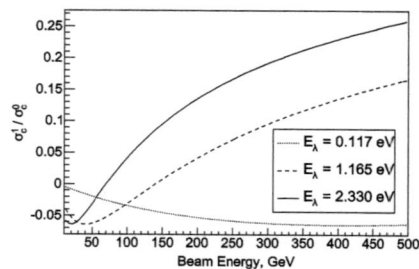

Figure 6.19: σ_c^1/σ_c^0 as a function of E_b for three laser energies.

6.6.2 Nonlinear Effects

When the density of the laser photons is very high, an electron can interact simultaneously with more than one laser photon [70–73]:

$$e + n\gamma \to e' + \gamma' , \, n \geq 1 \tag{6.29}$$

An intuitive picture in form of a Feynman diagram is given in Fig. 6.20.

This process should not be confused with multiple scattering. In the last case, an electron scatters first, afterward it travels further across the laser bunch and might scatter again. The electron in such multiple scattering processes is a real electron, whereas for nonlinear effects the electron is a virtual particle which is considered to absorb 'simultaneously' more than one photons.

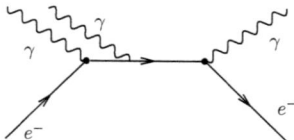

Figure 6.20: Feynman diagram of nonlinear Compton scattering with $n = 2$.

6.6 Potential Background Processes

For lasers with very high photon density, the electromagnetic field has an impact onto the behavior of the electron such that the electron is described by the 4-momentum q (quasi-momentum) as

$$q = p + \frac{m^2 \xi^2}{2p \cdot k_L} k_L, \qquad q^2 = (1+\xi^2)m^2, \tag{6.30}$$

where m is the electron mass, p the real 4-momentum, k_L the 4-momentum of the laser photon and ξ^2

$$\xi^2 = \frac{2n_\gamma r_e^2 \lambda_L}{\alpha}, \tag{6.31}$$

with n_γ as the local photon density in the laser pulse, r_e the classical electron radius, λ_L the laser wavelength and α the fine-structure constant. The important quantity ξ^2 characterizes the strength of the nonlinear effect and, as seen from its definition, it depends only on laser parameters. If the density of the laser pulse is sufficiently low, the quasi-momentum q becomes the usual expression for free particles.

The cross-section for nonlinear Compton scattering can be written as an incoherent sum of contributions [73]

$$\sigma_c = \sigma_1 + \sigma_2 + \sigma_3 \dots, \tag{6.32}$$

where

$$\sigma_n = f_{n1}(E_{beam}, \lambda_L, \xi^2, n) + P_e \lambda f_{n2}(E_{beam}, \lambda_L, \xi^2, n) \tag{6.33}$$

and λ the polarization of the laser light, P_e the polarization of the initial electrons and σ_n the probability for an electron to absorb n photons.

For each of the processes in (6.32), the maximum energy of the emitted photons is given by

$$\omega_{max}^n = \frac{nx}{(1 + nx + \xi^2)}, \tag{6.34}$$

where x is given by Eq. (6.2).

Three important properties can be derived from Eq. (6.34):

- absorption of more than one photon generates electrons with an energy below E_{edge};
- for the linear Compton scattering process (n=1), E_{edge} increases with growing ξ^2;
- ω_{max}^n is independent on the polarization of the initial state.

The consequences of nonlinear effects are illustrated in Fig. 6.21. Considering, for example, its right-hand side with $x = 4.8$ (as given by Eq. (6.2)), the curves represent (from right to left) the photon energy spectra for ξ^2=0, 0.1, 0.2, 0.3 and 0.5. For ξ^2=0, the spectrum corresponds to the Born approximation as given by (6.1), while the Compton edge becomes smaller with increasing ξ^2 and, at the same time, events with energy larger than the Compton edge appear. These events are contributions from nonlinear effects, with of $n > 1$ in Eq. (6.33).

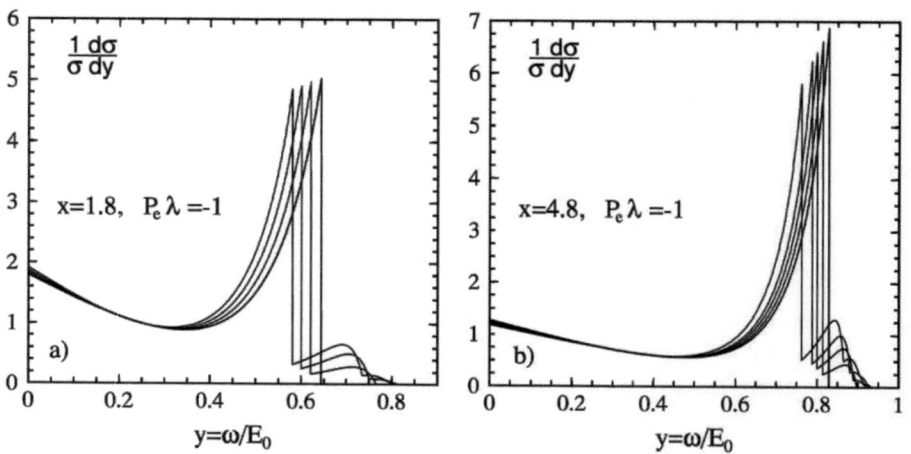

Figure 6.21: Normalized energy spectra of photons with different ξ^2 value. Left for $x = 1.8$ and right for $x = 4.8$, with x as given by Eq. (6.2). The curves (from right to left) correspond to $\xi^2 = 0$, 0.1, 0.2, 0.3, and 0.5 only for $x = 4.8$.

6.6.2.1 Impact of Nonlinear Effects

Assuming a laser bunch with Gaussian shape, the maximum laser density can be expressed as

$$n_\gamma^{max} = \frac{P_{laser}}{(2\pi)^{3/2}\sigma_x\sigma_y\sigma_z E_\lambda} = 1.80547 \cdot 10^{17} \text{photons/mm}^3 \ . \tag{6.35}$$

For a CO_2 laser with e.g. $P_{laser} = 4 \cdot 10^{-4}$ J (the power of the laser pulse) and $E_\lambda = 0.117$ eV, the maximum value of ξ^2 results in $4.16048 \cdot 10^{-6}$.

For such a ξ^2, the ratio of the cross-section of two photon absorption, σ_2, to the cross-section of one photon absorption, σ_1, is expected to be $\sigma_2/\sigma_1 < 4 \cdot 10^{-6}$. Since in general, ξ^2 depends on the position of the interaction in the laser bunch, ξ^2, respectively, σ_2/σ_1, is not a single number, but corresponds to a broad distribution. Figure 6.22 shows the distribution of σ_2/σ_1 for a Gaussian shaped CO_2 laser and unpolarized electrons.

Concerning the edge energy E_{edge}, we expect a shift of

$$E'_{edge} = \epsilon + E_\lambda - \omega^1_{max} \ , \tag{6.36}$$

with $\epsilon = E_b + (\xi^2 m^2)/(4E_b)$, the effective energy of the initial electron, and ω^1_{max} given by Eq. (6.34) for n=1.

For the particular CO_2 laser chosen the relative shift of the edge energy is then expected to be

$$\frac{E'_{edge} - E_{edge}}{E_{edge}} = 1.29 \cdot 10^{-6} \ . \tag{6.37}$$

6.6 Potential Background Processes

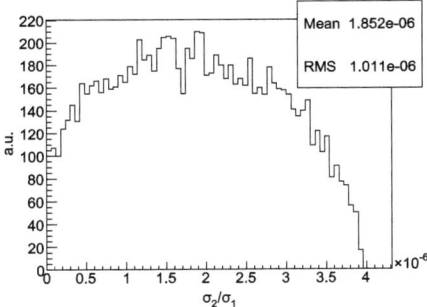

Figure 6.22: σ_2/σ_1 distribution for a Gaussian-like CO_2 laser bunch.

For lasers with smaller wavelength such as the infrared or green laser considered in the thesis, ξ^2 is even smaller, so that the impact of nonlinear effects becomes significantly smaller or negligible.

6.6.3 Breit-Wheeler Process

With an increase of the variable x defined in Eq. (6.2), e^+/e^- pair creation by high energy Compton photon collisions with laser photons leads to further background, which has the potential to disturb the edge electron behavior. The threshold of this reaction of $E_{\gamma,max}E_\lambda = m^2$ results in $x = 2(1+\sqrt{2}) \simeq 4.83$. If x is larger than 4.83, which happens when e.g. 250 GeV electrons collide with green laser light, associated $e^\pm$ pair background is generated (Breit-Wheeler process) [71, 72].

To evaluate the number of $e^\pm$ pairs a procedure similar to that used in Sect. 6.6.1.1 has been applied. Counting the number of $e^\pm$ pairs for values of the laser power P_{laser} between 10 and 60 Joule, a quadratic dependence $N_{e^+/e^-} = b_2 \cdot P_{laser}^2$ is obtained as shown in Fig. 6.23 (right) for a green laser at 500 GeV. Once the parameter b_2 is determined by a fit, it is possible to extrapolate N_{e^+/e^-} for lower laser pulse powers. A value of N_{e^+/e^-} of about 30 for $P_{laser} = 0.039$ J was thus found.

This evaluation was done with unpolarized electron beam. For the worst case of $\lambda P_e = -1$ and $E_b = 500$ GeV, the number of photons near the Compton edge is approximately doubled where the cross-section of the Breit-Wheeler process is larger resulting in $N_{e^+/e^-} \lesssim 60$ per bunch crossing.

All these values, including the last one, are considered to be unimportant and are expected not to affect a precise energy measurement.

6.6.4 Higher Order QED Corrections

The amplitude for a generic process in quantum field theory is calculated as a series of power of α, the fine-structure constant. This series is called the perturbation series. Often it is sufficient

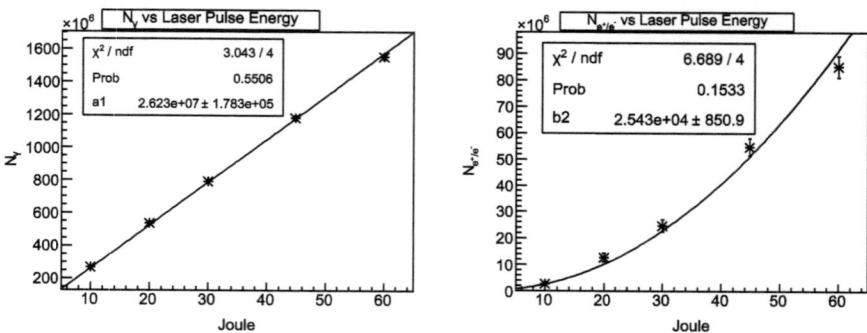

Figure 6.23: Left: Number of scattered photons vs. laser pulse power. Right: Number of e^+e^- pairs vs. laser pulse power.

to consider only the first term in the series, the Born approximation, which is proportional to α^2 and neglect the others, but for high energy beams, the term proportional to α^3 (virtual corrections) cannot be neglected since it might be few percent of the order-α^2 terms [74]. However, the higher order contributions have the same kinematics as the order-α^2 terms and since E_{edge} is defined only by the kinematics, no change is expected for E_{edge}.

Furthermore, besides the order-α^3 terms from the perturbation series, two other processes have to be considered, the double Compton process, $e + \gamma \rightarrow e + \gamma\gamma$, and the direct pair production reaction, $e^- + \gamma \rightarrow e^+e^+e^-$, since both cross-sections are of order-α^3. An example of Feynman diagrams for double Compton scattering and direct pair production is given in Fig. 6.24. It turns out that the soft part of the double Compton process is divergent towards zero photon energy. This divergence is as usual canceled against virtual corrections of the same order in α within a renormalization scheme.

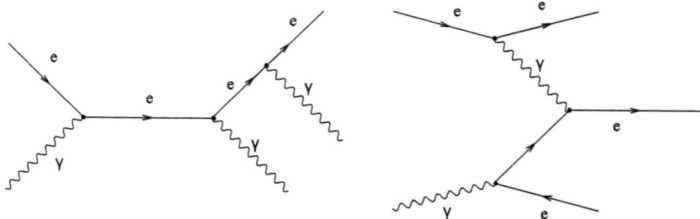

Figure 6.24: Example of Feynman diagrams for the processes $e+\gamma \rightarrow e+\gamma\gamma$ (left) and $e^-+\gamma \rightarrow e^+e^+e^-$ (right).

In the case of double Compton scattering, the final state electron also has a minimum energy which coincides with the Compton edge of the standard process $e + \gamma \rightarrow e + \gamma$. This can be understood as follows. The scattered electron has its minimum energy when the 2-photon system has its maximum energy. This happens when the two photons are collimated exactly in the forward direction. Their invariant mass is then zero and the two photons cannot be

6.7 Potential Systematic Error Sources

distinguished from a single photon. Hence, the situation is the same as for the usual Compton backscattering with only one photon in the final state.

The process $e^- + \gamma \to e^+e^+e^-$ has a production threshold as the Breit-Wheeler process. Considering a green laser, this threshold is near $E_b = 225$ GeV. At this energy, all final state particles are collimated exactly in forward direction and each particle carries one third of the initial beam energy, i.e. ~ 75 GeV. At 500 GeV, the final state electrons/positrons are characterized by a cut-off energy of 34.36 GeV, which is however far away from the fitting energy range of interest, namely of $E_{edge} \sim 26$ GeV or near by.

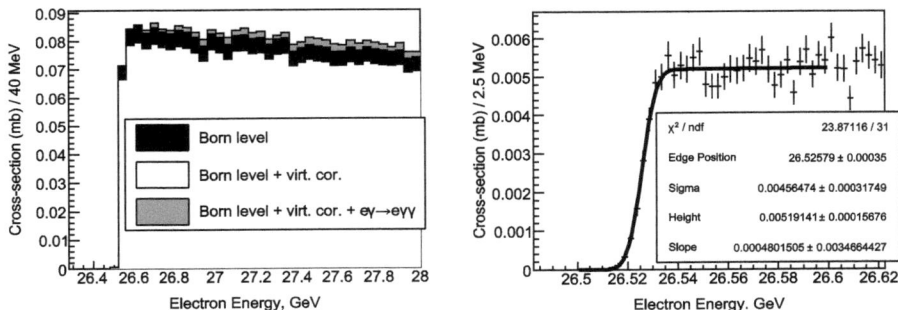

Figure 6.25: Left: Born cross-section of the Compton process (black histogram), Born plus order-α^3 cross-section (open histogram) and Born plus order-α^3 terms including the reaction $e + \gamma \to e + \gamma\gamma$ (shaded histogram) for $\lambda P_e = -1$ at 500 GeV. Right: Electron energy spectrum including all order-α^3 contributions and a beam spread of 0.3%. The line represents a fit with Eq. (6.18).

To confirm the expectations, the program COMRAD [75] was utilized to generate electron energy spectra including virtual corrections and the double Compton process. The left-hand side of Fig. 6.25 shows the distributions obtained near the Compton edge for a 500 GeV beam and a green laser with $\lambda P_e = -1$. Such conditions allow for largest higher order contributions. As can be seen, the Born spectrum is only enhanced by few percent, as stated above. Furthermore, an edge electron energy spectrum was generated including all order-α^3 contributions and a beam spread of 0.3% (right-hand side of Fig. 6.25) and using the ansatz (6.18), E_{edge} was determined. The fit result is $E_{edge} = 26.52579 \pm 0.00035$ GeV, which is only 1.5 standard deviation away from the Born expectation of 26.52525 GeV, i.e. the difference is smaller than 20 ppm. This example demonstrates that order-α^3 contributions have no measurable impact on beam energy measurements by Compton backscattering.

6.7 Potential Systematic Error Sources

Concerning the statistical error, it has been shown that with 10^6 Compton events per bunch crossing the accuracy on the beam energy $\Delta E_b/E_b$ can be brought below the requirement of

10^{-4}.

Different sources of potential systematic errors may affect the measurement of E_b and are discussed below. As outlined in Sect. 6.6, Compton processes beyond Born approximation such as multiple scattering, nonlinear effects in the $e \to \gamma$ conversion process, higher order QED corrections or e^+e^- pair creation will not significantly modify the scattered photon and edge electron behavior. As shown by simulation, their effect on $\Delta E_b/E_b$ is negligible even after summation over all non-Born approximations discussed.

6.7.1 Quartz Fiber Detector

The geometrical precision of quartz fibers is crucial for their precise assembly into plans and stacks. Arrangements of sufficient precisions including intrinsic fiber uncertainties are based on experience [76, 77] and can be kept to 2 μm or better for small devices. The position of the fibers can be accurately measured after the assembly and recorded in a database for use during analysis so that final arrangement errors may be less than few tenth of a micrometer. Also, any tilt of the detector with respect to the vertical direction is important. For example, a misalignment error of 1 mrad could result in an X_γ position shift up to 5 μm. Hence, the detector has to be aligned to better than 0.1 mrad in order to keep this bias contribution $\lesssim 0.2$ μm.

Possible errors caused by the fit procedure of the escaping $e^\pm$ position distribution have been checked by varying the fit region within reasonable values or by rebinning the spectrum or by omitting the first step of the two-step X_γ-fit procedure. There is no evidence found for a bias of the nominal fitted values, taking their statistical precision into account. Conservatively, we assign an error of 0.1 μm due to residual uncertainties from the fit.

The signal uniformity of the fiber sensor is also an important issue. Variations from fiber-to-fiber (or strip-to-strip) may have various reasons. There are statistical variations due to noise and fluctuations of the number of photons emitted, but also variations of the signal response laterally across the detector which are caused by fluctuations in the local properties of the sensor. Cerenkov light variations were already addressed in the GEANT simulations. The remaining fluctuations were studied by some additional Gaussian channel-to-channel signal variation of 0.5%, 1% and 2% of the total signal per fiber. It was found that the original position of interest is shifted by less than 1 μm for a response fluctuation not exceeding 1%. In practice, the level of uniformity across the sensor should be measured by an appropriate uniform illumination with particle beams and the individual channel response accounted for in the data analysis.

Imperfections within the fiber readout chain are difficult to estimate at the present stage of the project. However, in order to fulfill the request, we set a limit of 0.2 μm for fiber-to-fiber signal variation and instability.

6.7 Potential Systematic Error Sources

6.7.2 Avalanche Detector

Alternatively, recording the SR edge by means of the avalanche detector [51], the primary beam line position depends on the amount and shape of the fringe field of the magnet. If the 1% integrated B-field fraction for the fringe field as used in the simulation was varied between 0.5 and 2% and three field shapes are considered (a simple step function, a straight line between zero and the B-field strength and a Gaussian distribution) it was found that the edge positions were distributed over a range of 1.6 μm. Thereby, precise measurement of the fringe field is mandatory, in particular upstream of the magnet. We estimate a residual error of $\sim$0.2 μm for X_γ due to surviving uncertainties of the integrated B-field which includes imperfections of the fringe field. Also, additional errors of 0.1 and 0.2 μm due to imperfections in the fit procedure, respectively, electronics were assigned.

6.7.3 Beam and Laser Jitter

Since the beam and the laser widths are comparable in size, the laser has to be steered onto the electrons such that both beams collide centrally. Otherwise, a shift of the center-of-gravity of the scattered photons is generated. Options for laser spot size monitoring and its stabilization are discussed in [13]. Electron position and emittance are supplied by BPMs, respectively, wire-scanner systems distributed within the BDS. Beam jitter studies suggest for $\sigma_{jitter} = (0.1 \div 0.5) \cdot \sigma_{x(y)}$ [9, 78], where $\sigma_{x(y)}$ is the bunch size in X(Y)-direction[21]. For a beam extension $\sigma_x = 20$ μm for example, the horizontal jitter is small, in the order of few micrometer, and negligible in the vertical direction. If one restricts any shift of X_γ to be less than 0.3 μm, constraints for the distance between laser and electron beam centers as a function of the laser spot size can be derived. With $\sigma_x = 20\mu$m and a laser spot size of $\sim$150 μm, the position of the laser has to be stable within 12 μm. Larger spot sizes relax this condition, whereas a bigger electron bunch size aggravates the condition considerably. For example, a 50 μm bunch requires a laser spot of the order of 300 μm and a laser jitter of less than 10 μm in order to maintain the photon centroid shift below 0.3 μm. Luminosity loss due to larger laser spot sizes can be compensated by either an increase of the laser power or an increased pulse length, or a combination of both. If the pulse duration is substantially increased it seems of advantage to consider horizontal instead of vertical beam crossing. In conclusion, in order to design a laser system which restricts the shift of X_γ due to non-central collisions of electron and laser pulses to less than $\sim$0.3 μm, some R&D effort is needed.

6.7.4 X_γ Determination

Summing all contributions quadratically, the total error associated to the γ-ray centroid position is $\Delta X_\gamma \simeq 0.8$ μm for the quartz fiber detector-converter system, while the SR edge approach

---

[21] Some machine experts prefer to use the smaller number. The size of the jitter will depend on the stability of the ILC beam line components, on energy and kicker jitter and on the performance of train-to-train and intra-train feedback.

provides ~ 0.6 μm. Both uncertainties are smaller than the required figure of ~ 1 μm.

6.7.5 X_{edge} Determination

Concerning the measurement of the electron endpoint, we found for X_{edge} an uncertainty of about 4 μm for the CO_2 laser, while the infrared (green) laser provides values of 6 and 7 (12 and 14) μm for the DSD, respectively, QFD detector. The differences in precision are mainly due to different event rates per detector strip or fiber.

For the diamond strip detector we assume a similar alignment precision as for the X_γ fiber detector discussed above and a bias estimate of 1.5 μm due to imperfections of the detector response.

The yield of the Cerenkov light in quartz fibers varies considerably in the vicinity of the Compton edge. Here, the number of incident electrons per fiber ranges from $\sim$150 to only a few or zero. Correspondingly, the number of photoelectrons in the light signal detector also varies considerably, which in turn requires high quantum efficiency at the wavelength of maximum scintillation and excellent single-photon detection capability. If we e.g. assume a zero-signal for fibers with less than 10 incident electrons, the refitted endpoint positions were found to be within ± 0.6 μm compared to the original values. This suggests to assign a total uncertainty associated to detector effects of 2 μm.

6.7.6 Method A

Relying on method A (Sect. 6.3.2), $\Delta E_b/E_b$ is controlled by the accuracy of the integrated B-field of the spectrometer magnet, $\Delta B/B$, the drift distance to the detector plane, $\Delta L/L$, and the offset of the edge position with respect to the primary beam line, $\Delta D/D$, see Eq. (6.8) and Fig. 6.6. The drift distance can be precisely monitored using an interferometer [79]. For an accuracy of $\Delta L \simeq 100$ μm, which is feasible, the relative error of L becomes few times 10^{-6} and hence negligible. The required accuracy for the distance between the undeflected beam and the endpoint in the order of few micrometers is only possible when a CO_2 laser is used. To achieve such a precision the X_γ and X_{edge} detectors should be installed on a common frame and rigidly connected in order to avoid relative position movements. In this way and with a frame made out of a material with a small expansion coefficient like carbon the relative distance error between both devices can be kept below 1 μm, even with a $\pm 5^o$ change in tunnel temperature. More important contributions to $\Delta D/D$ constitute the uncertainties of the X_γ and X_{edge} position measurements themself.

As emphasized in Sect. 6.3.2, the relative error of the B-field integral should be close to $2 \cdot 10^{-5}$ in order to reach the required beam energy precision. Aspects necessary to fulfill this challenging request are summarized in [37]. Here we point out that, independent of the endpoint detector utilized, in addition to the bending field provided by the spectrometer dipole itself, several other sources of magnetic fields may be present in the ILC tunnel which might influence the path of the electrons. A large effect can come from the earth's field and other

6.7 Potential Systematic Error Sources

contributions might arise from cables which provide current for magnets. Such fields within the space between the Compton IP and the detector plane could spoil the endpoint position measurement, even if this space is free of any magnetic element. The effect of e.g. the earth's field if normal to the full edge electron trajectory will shift the impact point 25 m downstream of the magnet by approximately 12 μm. Hence, the ambient field can be critical and should be either shielded or measured by e.g. a fluxgate magnetometer. Such an instrument allows to monitor any variation of the ambient magnetic field with time and endpoint corrections should be applied. It is estimated that such a field has to be known with a relative accuracy of (better than) 10% to ensure a tolerable contribution of $\lesssim 1.5$ μm to the overall X_{edge} uncertainty.

Considering all the arguments proposed, an error for X_{edge} of 4.9 μm is obtained while for X_γ the error has been evaluated to be about 0.6 μm. The error on the displacement D, $\Delta D = \sqrt{\Delta X_\gamma^2 + \Delta X_{edge}^2}$, is thus about 4.9 μm, a value still acceptable.

6.7.7 Method B

Adding all uncertainties together, the total error of X_{edge} can be expected to be close to 6.6 or 7.3 μm (12.2 or 14.2 μm) if an infrared (green) laser is used in the spectrometer. The dominating fraction of the error comes from statistics so that larger data samples would decrease these uncertainties. In general, all estimated uncertainties are very close to or less than the errors anticipated in Sects. 6.3.3. The error on X_γ measurement was found to be about 0.6 μm, below the requirement of 1 μm.

If method B will be realized, precise position of the unscattered beam, X_{beam}, at the detector plane is also required. Cavity beam position monitors with single-bunch resolution of few hundred (or less) nanometer are best suited. To be conservative, we assume an error for X_{beam} of 1 μm which has to be added in quadrature with the uncertainty from possible charged particle background expected for one of the proposed spectrometer locations (Sect. 6.8). In the worst case, the total uncertainty of X_{beam} results in ~ 1.2 μm which is well within the requirement.

In addition, if the B-field integrals for the endpoint and beam electrons are different, $(\int Bdl)_{edge} \neq (\int Bdl)_{beam}$, the expression for the beam energy (6.16) must be rewritten as

$$E_b \propto \frac{R(X_{edge} - X_\gamma) - (X_{beam} - X_\gamma)}{(X_{beam} - X_\gamma)}, \tag{6.38}$$

with

$$R = \frac{(\int Bdl)_{beam}}{(\int Bdl)_{edge}}. \tag{6.39}$$

Hence, the error for the beam energy as a function of the relative uncertainty of R is

$$\frac{\Delta E_b}{E_b} = \frac{(X_{edge} - X_\gamma)}{(X_{edge} - X_{beam})} \frac{\Delta R}{R}, \tag{6.40}$$

where the approximation $R \approx 1$ has been implied. If the corresponding particle positions 25 m downstream of the spectrometer magnet are taken into account, the ratio $(X_{edge} - X_\gamma)/(X_{edge} - $

$X_{beam}) = 1.2$ (1.1) for the infrared (green) laser. This means that for $\Delta R/R \simeq 5 \cdot 10^{-5}$ or better and any value of R different from 1, Eq. (6.16) is needed to be modified as indicated above. If R equals 1 (within few times 10^{-5}), no correction has to be applied. With today's common B-field and $\int Bdl$ measurement techniques such precision for R can be achieved without too much efforts.

Throughout the thesis, the approximation for the bend of an electron passing a B-field in Eq. (2.19) has been used and higher order contributions as given by Eq. (2.18) were neglected. Figure 6.26 shows the ratio between the first and second term of the Taylor expansion (2.18)

$$\frac{\frac{L}{2} \cdot \left(\frac{l}{R}\right)^3}{\left(L + \frac{l}{2}\right) \cdot \frac{l}{R}}$$

against the beam energy for three different lasers.

As can be seen, for a green laser ($E_\lambda = 2.33$ eV) and $E_b = 500$ GeV, the second term is largest and becomes 10^{-3} compared to the first one. This means that this term cannot be neglected. It can, however, be easily calculated with an accuracy of 0.1% and taken into account in the analysis without the need of a precise theory.

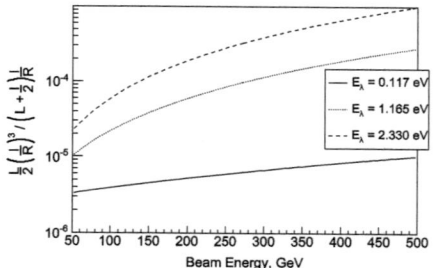

Figure 6.26: Ratio between the first and second term of the Taylor expansion (2.18) as function of the beam energy for three laser frequencies.

Of the same order of magnitude is the correction due to emission of synchrotron radiation by the electrons when passing the spectrometer magnet [80].

Basically, whatever will be the final choice for the electron detector more elaborated simulation studies are mandatory. In particular, physics processes in the sensor material, basic parameters of the associated electronics and backgrounds need to be included in Monte Carlo studies. Such details may affect the edge position and its shape and could limit the performance of the spectrometer. Studies of this kind are, however, beyond the scope of this thesis.

The idea to pulse the laser on every ILC bunch may be diluted for background studies. If e.g. the laser is pulsed on nine out of ten bunches, every 10th pulse can be used for background informations.

6.8 Suitable Energy Spectrometer Locations

Although today's beam delivery system [58] will be further developed within the next years, basic properties are not expected to be modified. We propose three alternatives for possible locations of the Compton spectrometer within the BDS, while keeping major design parameters of the spectrometer unaltered. Each of the proposals has pros and cons and the spectrometer viability requires sometimes, depending on the location, slight modifications of the present BDS. An overall view of the BDS is shown in Fig. 6.27, where also potential locations for the Compton spectrometer are indicated.

Common to all alternatives is the demand to locate the spectrometer upstream of the energy collimation system[22] to avoid significant muon background excess relative to the rate from normal collimation losses.

The straight-forward approach suggests to locate the spectrometer in an existing free-space region of the BDS. The amount of space needed is determined by the drift distance of at least 25 m to the detector system, the length of the magnet of 3 m and the 6 m long vacuum chamber upstream of the dipole in which the Compton IP is contained. The sum of these components of 35 m has to be enlarged by additional space to accommodate two ancillary magnets with corresponding drift regions to compensate the bend of the spectrometer magnet. Hence, in total $60 \div 70$ m free space is needed[23]. Far upstream of the physics e^+e^--IP such free space of some 65 m exists, see Fig. 6.27. The transverse dimensions of the beam at the Compton IP of about 20 μm versus 2 μm perfectly match the expected spot size of the laser. Additional muon background generated by backscattered electron interactions further downstream was estimated and would only increase the muon rate by a small amount [81], independent of the laser wavelength. This suggestion locates the spectrometer on a direct line of sight to the main linac, which means that backgrounds in this region are likely to be significant. In particular, charged particles off in energy may affect the position of the beam in the cavity BPM. Cavity beam position monitors measure the centroid of the particle's charge distribution and, hence, particles with less energy than E_b are stronger deflected by the spectrometer magnet and could shift the measured beam position. Halo and tail generation estimates based on simulation [82] reveal that at the exit of the linac the beam profile is superimposed by a symmetric halo extending to about ± 300 (50) μm in X(Y)-direction[24]. The fraction of particles off in energy was estimated to be few times 10^{-5} with a broad energy spectrum that sharply peaks very close to the nominal beam energy. A simple tracking procedure up to the BPM installed 25 m from the spectrometer magnet indicates a shift of X_{beam} of 0.65 μm which we consider of not being

---

[22]The energy collimation system performs efficient removal of halo particles which lie outside the acceptable range of energy spread.

[23]It would be very helpful if in any new BDS design a suitable spectrometer dipole is a priori foreseen as a standard BDS magnet. This would substantially relax space (and other) requirements for the Compton spectrometer.

[24] The electron bunches at the end of the linac were found to be very well described by pure Gaussian distributions with horizontal and vertical dimensions of $\sigma_x = 39.0$ μm and $\sigma_y = 1.80$ μm, respectively.

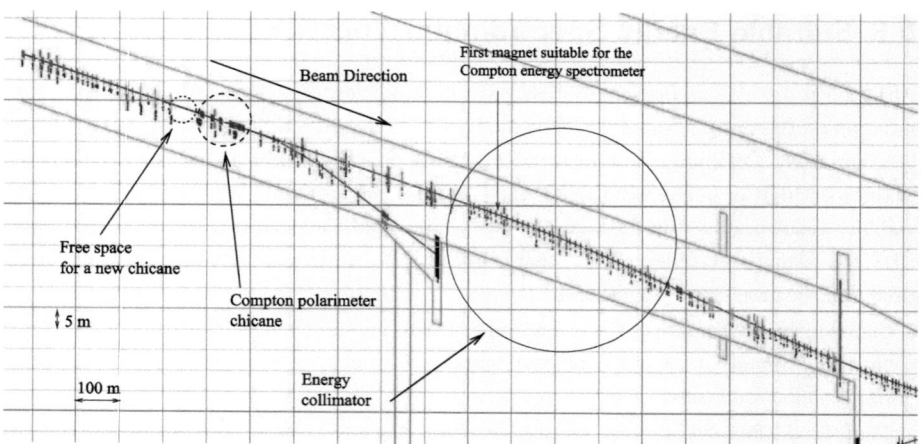

Figure 6.27: General view of the beam delivery system with possible locations of the Compton spectrometer.

catastrophic. It is also estimated that this background does not affect the Compton endpoint position in a significant manner. The synchrotron photons from the quadrupole fields within the linac and the beginning of the BDS have a $\sim 10^2$ times lower critical energy [82–84] than those from the spectrometer magnet and are considered to be of no serious issue. However, due to the uncertainties in the charged particle background simulation it is favorable to locate the spectrometer after a protective bend, so that the beam position will be much less impacted.

A major constraint for the design of the Compton spectrometer is the synchrotron radiation emittance dilution from the additional spectrometer magnets. Employing the magnet as discussed in [13] and similar ancillary magnets, an emittance growth of about 0.5% at 250 GeV is expected, which might be considered as acceptable. Since the emittance scales with the sixth power of the beam energy, further studies have to reveal whether emittance dilution at 500 GeV beam energy can be tolerated.

A second option for the spectrometer location consists in employing one or more magnets of the present BDS as the Compton spectrometer dipole. Since, however, an individual magnet with desirable properties does not exist, we suggest to combine several consecutive bending magnets. At the beginning of the energy collimation section directly after the first magnet, see Fig. 6.27, such magnets[25] might be combined to provide the desired bending power. In particular, if the laser IP is located about 3 m upstream of magnet 1, a combination of the following six magnets (magnet 1, ..., magnet 6) provides sufficient particle separation. For example, separation between the backscattered γ-rays and the beam line results in 18 mm after passing magnet 6, while the distance of the beam to the edge electrons is 26 mm for a CO_2 and 98 mm for an 1.165 eV infrared laser. Thus, by locating the detector system close to magnet 7 convenient measurements of the positions of the Compton recoil particles and the beam line

---

[25]Each magnet has a B-field of 291.68 Gauss, a length of 2.4 m and space in between of 12.3 m.

can be performed. The transverse beam profile at the laser IP is sufficiently small so that the beam is completely covered by the laser spot. Additional muon background from Compton electrons is tolerable since many of these electrons will hit either closely located magnets or spoilers of the energy collimation system [81]. This option also allows to insert the laser light into the vacuum pipe downstream of magnet 1 which makes strict head-on collisions with the beam possible.

However, the horizontal aperture of the magnets has to be continuously increased towards the bending direction so that the edge electrons pass in B-fields with properties as demanded. In particular, at the exit of magnet 6 the vacuum chamber has to have a horizontal aperture of 115 mm for infrared laser light scattering. Furthermore, if method A is employed for beam energy measurements, the B-field integral over all six magnets has to be known with 20 ppm precision. Or, for method B, the three-point measurement approach, sufficient field uniformity within the bending plane up to $X = 115$ mm has to be ensured. Whatever method for E_b determination will be realized, the demands for the magnet system are challenging. This alternative for the spectrometer location is advantageous since no additional magnets are needed and, thereby, further growth of the beam emittance is a priori avoided.

The third alternative for a location of the Compton spectrometer consists in employing the magnetic chicane proposed for high energy polarization measurements [85]. In particular, the 4-magnet polarimeter chicane with the laser IP in the mid-point is supposed to be supplemented by the position sensitive detector system, which should be located upstream but close to the fourth magnet. Also, some dedicated adjustments of space, laser and detector conditions are needed to ensure polarization and beam energy measurements simultaneously with precisions as anticipated. However, the present baseline polarimeter design aims to operate the chicane with constant field settings over a large range of beam energies, while the Compton based beam energy spectrometer intends to adjust the B-field to a constant bending power of e.g. 1 mrad. Whether both approaches can be merged to a common proposal requires detailed studies. Possible muon background increase from Compton electron interactions was estimated to be tolerable [81]. It is also obvious that additional dilution of the beam emittance caused by such E_b measurements is ruled out.

6.9 Summary

A novel, non-invasive method of measuring the incident beam energy, E_b, at the International e^+e^- Linear Collider is proposed. Laser light scatters head-on off ILC bunches and generates Compton electrons and photons. After the Compton IP, the scattered particles as well as the non-interacting beam electrons (99.9995% of them) pass through a dipole magnet so that further downstream access to each particle type is possible. E_b measurements can be performed continuously on a bunch-to-bunch basis while the electron and positron beams are in collision.

One approach to infer E_b, method A, relies on the beam energy dependence of the momentum of the scattered electrons at the kinematic endpoint, the edge energy. Combining the B-field integral of the dipole with the position of the edge electrons relative to the incident beam provides the energy of the edge electrons and, thereby, E_b. However, integrated field uncertainties close to $2 \cdot 10^{-5}$ and position measurements with an accuracy of at least few micrometers are required to achieve the anticipated value of 10^{-4}. The last demand is very challenging and is mainly the reason to follow a different approach, denoted as method B. By measuring three particle positions, the position of the Compton scattered γ-rays, X_γ, the position of the edge electrons, X_{edge}, and that of the beam, X_{beam}, downstream of the spectrometer magnet allows to deduce E_b with precisions of 10^{-4} or better. Such precisions, however, require to measure the distance $X_{edge} - X_{beam}$ with an accuracy of about ten micrometer and X_γ with $1 \div 2$ μm uncertainty. Both requirements seem to be achievable. In particular, the distance $X_{edge} - X_{beam}$ is beam energy independent and accumulation over many bunches decreases its statistical error substantially.

It has been shown that effects beyond the Born approximation in the laser crossing region are very small. They only lead to a negligible shift of the edge electron position, X_{edge}.

Geometrical constraints and acceptable emittance dilution of beam particles when passing the dipole magnet require a spectrometer length of at least 30 m. The geometrical constraints in conjunction with free space options within the present beam delivery system preclude the usage of a CO_2 laser, while an infrared (with $E_\lambda = 1.165$ eV) or a green laser (with $E_\lambda = 2.33$ eV) are both suitable. To achieve e.g. 10^6 Compton events per bunch crossing, a pulse power of 30 mJ, respectively, 24 mJ with a pattern that matches the pulse and bunch structure at the ILC is needed. Such lasers are presently commercially not available, but R&D is ongoing within the ILC and other communities.

For particle position measurements, detectors with high spatial resolution have to be pursued. As a promising option for edge electron and γ-ray center-of-gravity measurements quartz fiber detectors are suggested because they are very radiation hard and ultrafast. An alternative to the X_γ quartz fiber detector (in conjunction with e.g. a 16 radiation length tungsten converter) consists in measuring the edge position of synchrotron radiation light generated by beam particles when passing the spectrometer magnet, as discussed in [51]. A device based on gas amplification was considered in more details and simulations demonstrated its reliability for our purpose. The position of the non-interacting beam particles needs to be known with micrometer accuracy which can be relative easily achieved by modern cavity beam position monitors.

The method proposed to perform energy measurements of the incident beam at the ILC is thought to be a complementary and cross-check approach to the canonical concept of a BPM-based energy spectrometer. Both methods intend to achieve a precision of 10^{-4} on a bunch-to-bunch basis. The method studied in this thesis seems to accomplish the objective, but more detailed studies are mandatory and a prove-of-principle experiment [86] should to be performed to test the three-position measurement approach.

The experience at SLC and LEP proved that independent measurements of the beam energy are important. Both the Compton and the BPM-based spectrometers are designed to provide

6.9 Summary

an absolute measurement of the beam energy with a relative accuracy of 10^{-4}. Cross-calibration of the spectrometers would provide an important and valuable control of their systematic errors. Also, energy measurements at the Z-pole would provide a unique possibility for an early calibration in a well understood physics regime. Although Z-pole calibration measurements are not part of the current ILC baseline design [9], it is argued [87] that the baseline should be modified to include such reference. In addition, physics reference channels, such as $e^+e^- \to \mu^+\mu^-\gamma$ where the muons are resonant with the known Z-mass, are foreseen to provide valuable checks of the collision energy scale, but only long after the data were recorded.

Conclusions

For the physics program at the ILC the center-of-mass energy $\sqrt{s}$ has to be controlled with excellent accuracy. The basis for the knowledge of the luminosity-weighted $\sqrt{s}$ for physics analyses is provided by measurements of the beam energy upstream of the e^+/e^- interaction point with a precision of 10^{-4} or better. The present scheme at the ILC foresees the usage of a spectrometer composed of a 4-magnet chicane with beam position monitors (BPMs) (see Figs. 3.2, 3.5b and 5.5).

A prototype of such a device was commissioned in End Station A (ESA) at the Stanford Linear Accelerator Center (SLAC) in 2006/2007. The goal of the experiment was to study its performance and reliability to gain experience for the planning of future spectrometers.

To determine the beam energy, monitoring the B-field integral of the magnets is very important besides the measurement of the transverse position of the beam upstream, downstream and in the mid-chicane. For monitoring the B-field integral, an accuracy of $5 \cdot 10^{-5}$ or better was demanded. In the thesis, measurements of the dipole magnets performed at the SLAC laboratory are reported which is needed to understand their properties and to verify an appropriate procedure to measure the B-field integral during data taking runs. A relative integrated B-field precision of $18.4 \cdot 10^{-5}$ was found, with major contributions caused by alignment errors. Also several suggestions to improve substantially this accuracy are reported.

The resolution of relative beam energy measurements is also evaluated. To determine the dipole-induced displacement of the beam in the mid-chicane, it was necessary to measure two positions of the beam, the position of the deflected beam in between the second and third magnet and that of the undeflected beam, respectively, the position of the extrapolated beam trajectory at the same beam line position (dashed line in Fig. 3.2). The offset or dispersion is determined from a BPM installed in the mid-chicane and the extrapolated position by using informations from the BPMs upstream and downstream of the chicane. To be able to use the data from the BPMs downstream of the chicane, a necessary prerequisite of the spectrometer is to restore downstream the upstream beam. In other words, the chicane has to "handle" the beam in a symmetric manner. This condition was not fulfilled at ESA. Thereby, the BPMs downstream could not be integrated into the analysis which in turn provides a worse beam energy resolution. At the end, the relative resolution of the beam energy was found to be $8.5 \cdot 10^{-4}$.

Experiences at SLC and LEP proved that complementary measurements of the beam energy are necessary in order to cross-check and cross-calibrate the results from the BPM-based spectrometer. In the thesis, a novel, non-destructive method for beam energy monitoring using

Compton backscattering of laser light on beam particles is studied. Previous experiments performed at the storage rings BESSY and VEPP-4M could not be copied for a bunch related beam energy determination at the ILC. Therefore, the method proposed differs in many respects and can be summarized as follows. After crossing the electron beam with a laser, a dipole magnet separates the backscattered undeflected photons, the unscattered beam particles and, with larger angle, the Compton scattered electrons. Downstream of the magnet the position of the scattered electrons with smallest energy and largest bend angle (edge position) and the center-of-gravity of the backscattered photons are measured. Combining these informations with the B-field integral (method A) or with the position of the unscattered beam (method B), the beam energy can be inferred. Both methods are studied in details. It was found that for the first method a laser with large wavelength is preferable, whereas the second approach requires lasers with short wavelengths. Detailed simulations are performed to evaluate possible detection options. For the Compton scattered electrons, a diamond sensor or quartz fiber detector seems to be suitable, whereas for the photons a quartz fiber or a novel avalanche detector of Ref. [51] is the best choice. Potential background processes within the laser-electron interaction region are evaluated and possible impacts on beam energy measurements are found to be negligible. It has been shown that with 10^6 Compton events per bunch crossing the statistical error as well as potential systematic errors can be brought well below the requirement of $\Delta E_b/E_b = 10^{-4}$. The method proposed is found to be very promising and accomplishes the objective at the ILC, but needs experimental verification.

Bibliography

[1] LEP/SPS Home Page, http://sl-div.web.cern.ch/sl-div/.

[2] G. Arnison et al. Experimental observation of isolated large transverse energy electrons with associated missing energy at $\sqrt{s} = 540$ GeV. *Phys. Lett.*, B122:103–116, 1983.

[3] G. Arnison et al. Experimental observation of lepton pairs of invariant mass around 95 GeV/c^2 at the CERN SPS collider. *Phys. Lett.*, B126:398–410, 1983.

[4] S. W. Herb et al. Observation of a dimuon resonance at 9.5-GeV in 400-GeV proton - nucleus collisions. *Phys. Rev. Lett.*, 39:252–255, 1977.

[5] S. Abachi et al. Search for high mass top quark production in $p\bar{p}$ collisions at $\sqrt{s} = 1.8$ TeV. *Phys. Rev. Lett.*, 74:2422–2426, 1995.

[6] F. Abe et al. Observation of top quark production in $\bar{p}p$ collisions. *Phys. Rev. Lett.*, 74: 2626–2631, 1995.

[7] B. Barish. Linear Collider Technology Recommendation, Presented at ILCSC/ICFA Special Meeting IHEP, Beijing, China, Aug 19 2004.

[8] J. E. Augustin et al., Linear Collider, Final International Technology Recommendation Panel Report.

[9] J. Brau et al. International Linear Collider Reference Design Report. 1: Executive summary. 2: Physics at the ILC. 3: Accelerator. 4: Detectors, ILC-REPORT-2007-001.

[10] M. Hildreth et al. Linear Collider - BPM-based energy spectrometer. http://www-project.slac.stanford.edu/ilc/testfac/ESA/projects/T-474.html, 2004.

[11] M. Woods et al. A test facility for the International Linear Collider at SLAC End Station A, for prototypes of beam delivery and IR components. arXiv:physics/0505171, 2005.

[12] M. Woods et al. Test beam studies at SLAC End Station A, for the International Linear Collider, Contributed to European Particle Accelerator Conference (EPAC 06), Edinburgh, Scotland, Jun 26-30 2006.

[13] N. Muchnoi, H. J. Schreiber and M. Viti. ILC beam energy measurement by means of laser Compton backscattering. *Nucl. Instrum. Meth.*, A607:340–366, 2009.

[14] P. Achard et al. Measurement of the mass and the width of the W boson at LEP. *Eur. Phys. J.*, C45:569–587, 2006.

[15] The ALEPH, DELPHI, L3, OPAL, SLD Collaborations, the LEP Electroweak Working Group, the SLD Electroweak and Heavy Flavour Groups. Precision electroweak measurements on the Z resonance. *Phys. Rept.*, 427:257, 2006.

[16] S. T. Boogert and D. J. Miller. Questions about the measurement of the e+ e- luminosity spectrum. arXiv:hep-ex/0211021, 2002.

[17] K. Mönig. Measurement of the differential luminosity using Bhabha events in the forward tracking region at TESLA. LC-PHSM-2000-060, 2000.

[18] P. Garcia-Abia, W. Lohmann and A. Raspereza. Prospects for the measurement of the Higgs boson mass with a linear e+ e- collider. *Eur. Phys. J.*, C44:481–488, 2005.

[19] A. Raspereza. Beam related systematics in Higgs boson mass measurement. arXiv:hep-ex/0412049, 2004.

[20] A. A. Sokolov and I. M. Ternov. On polarization and spin effects in the theory of synchrotron radiation. *Sov. Phys. Dokl.*, 8:1203–1205, 1964.

[21] V. Bargmann, L. Michel and V. L. Telegdi. Precession of the polarization of particles moving in a homogeneous electromagnetic field. *Phys. Rev. Lett.*, 2:435, 1959.

[22] A. C. Melissinos. Energy measurement by resonant depolarization, Presented at CERN Accelerator School: Course on Advanced Accelerator Physics, Rhodes, Greece, Sep 20 - Oct 1 1993.

[23] L. Arnaudon et al. Accurate determination of the LEP beam energy by resonant depolarization. *Z. Phys.*, C66:45–62, 1995.

[24] R. Assmann et al. Spin dynamics in LEP with 40-100 GeV beams, Prepared for 14th International Spin Physics Symposium (SPIN 2000), Osaka, Japan, Oct 16-21 2000.

[25] R. Assmann et al. Calibration of centre-of-mass energies at LEP2 for a precise measurement of the W boson mass. *Eur. Phys. J.*, C39:253–292, 2005.

[26] A. H. Compton. A quantum theory of the scattering of X-rays by light elements. *Phys. Rev.*, 21:483–502, 1923.

[27] A. Hinze and K. Mönig. Measuring the beam energy with radiative return events. arXiv:physics/0506115, 2005.

[28] R. Klein et al. Beam diagnostics at the BESSY I electron storage ring with Compton backscattered laser photons: Measurement of the electron energy and related quantities. *Nucl. Instrum. Meth.*, A384:293–298, 1997.

[29] R. Klein et al. Measurement of the BESSY II electron beam energy by Compton backscattering of laser photons. *Nucl. Instrum. Meth.*, A486:545–551, 2002.

[30] A. G. Shamov et al. Tau mass measurement at KEDR. *Nucl. Phys. Proc. Suppl.*, 181-182: 311–313, 2008.

[31] V. E. Blinov et al. Review of beam energy measurements at VEPP-4M collider: KEDR/VEPP-4M. *Nucl. Instrum. Meth.*, A598:23–30, 2009.

[32] J. Kent et al., Presented at IEEE Particle Accelerator Conf., Chicago, Ill., Mar 20-23, 1989.

[33] M. E. Levi, J. Nash, and S. Watson. Precision measurements of the SLC spectrometer magnets. *Nucl. Instrum. Meth.*, A281:265–276, 1989.

[34] M. E. Levi et al., Presented at IEEE Particle Accelerator Conf., Chicago, Ill., Mar 20-23, 1989.

[35] J. Kent et al. Design of a wire imaging synchrotron radiation detector, Presented at IEEE 1989 Nuclear Science Symposium, San Francisco, CA, Jan 15-19, 1990.

[36] R. Chritin et al. Determination of the bending field integral of the LEP spectrometer dipole. *Nucl. Instrum. Meth.*, A545:31–44, 2005.

[37] V. N. Duginov et al. The beam energy spectrometer at the International Linear Collider. LC-DET-2004-031, 2004.

[38] B. I. Grishanov et al. ATF2 proposal. SLAC-R-771, 2005.

[39] S. Walston et al. Performance of a high resolution cavity beam position monitor system. *Nucl. Instrum. Meth.*, A578:1–22, 2007.

[40] M. Slater et al. Cavity BPM system tests for the ILC energy spectrometer. *Nucl. Instrum. Meth.*, A592:201–217, 2008.

[41] R. Lorenz. Cavity beam position monitors, Contributed to 8th Beam Instrumentation Workshop (BIW 98), Stanford, CA, May 4-7 1998.

[42] P. B. Wilson. Introduction to wake fields and wake potentials. *AIP Conf. Proc.*, 184: 525–564, 1989.

[43] P. B. Wilson. High-energy electron linacs: applications to storage ring rf systems and linear colliders. *AIP Conf. Proc.*, 87:450–555, 1982.

[44] A. Lyapin. Strahllagemonitor für das TESLA-Energiespektrometer, Dissertation. TU-Berlin, Berlin, 2003.

[45] H. J. Schreiber et al. Magnetic measurements and simulations of a 4 magnet dipole chicane for the International Linear Collider, Presented at Particle Accelerator Conference PAC07, Albuquerque, New Mexico, Jun 25-29 2007.

[46] M. Viti and S. Kostromin. Magnetic measurements for magnets 10D37. ILC-SLACESA TN-2008-1, 2008.

[47] N. Morozov. Magnetic field simulation for the 10D37 magnet. ILC-SLACESA TN-2006-2, 2006.

[48] METROLAB Instruments SA, PT 2025 NMR Teslameter.

[49] CERN Accelerator School Proceedings, 1992 and 1998.

[50] V. Duginov and S. Kostromin. Measurements of residual magnetic fields near magnets and along the chicane beamline in End Station A. ILC-SLACESA TN-2006-5, 2006.

[51] K. Hiller et al. ILC beam energy measurement based on synchrotron radiation from a magnetic spectrometer. *Nucl. Instrum. Meth.*, A580:1191–1200, 2007.

[52] V. Gharibyan, N. Meyners and P. Schuler. The TESLA Compton polarimeter. LC-DET-2001-047, 2001.

[53] E. Feenberg and H. Primakoff. Interaction of cosmic-ray primaries with sunlight and starlight. *Phys. Rev.*, 73(5):449–469, 1948.

[54] H.J. Schreiber. International Linear Collider Workshop, Valencia, Spain, Nov 6-10 2006.

[55] T. Omori et al. Design of a polarized positron source for linear colliders. *Nucl. Instrum. Meth.*, A500:232–252, 2003.

[56] S. Schreiber et al. Running experience with the laser system for the RF gun based injector at the TESLA Test Facility linac. *Nucl. Instrum. Meth.*, A445:427–431, 2000.

[57] M. Ross. Laser-based profile monitor for electron beams, Presented at Particle Accelerator Conference (PAC 03), Portland, Oregon, May 12-16 2003.

[58] A. Seryi et al. Design of the Beam Delivery System for the International Linear Collider, Presented at Particle Accelerator Conference (PAC 07), Albuquerque, New Mexico, Jun 25-29 2007.

[59] N. Muchnoi. ILC beam energy measurement using Compton backcattering, In the Proceedings of 2007 International Linear Collider Workshop (LCWS07 and ILC07), Hamburg, Germany, May 30 - Jun 3 2007.

[60] W. Adam et al. Radiation hard diamond sensors for future tracking applications. *Nucl. Instrum. Meth.*, A565:278–283, 2006.

Bibliography

[61] P. Gorodetzky et al. Quartz fiber calorimetry. *Nucl. Instrum. Meth.*, A361:161–179, 1995.

[62] S. Agostinelli et al. GEANT4: A simulation toolkit. *Nucl. Instrum. Meth.*, A506:250–303, 2003.

[63] S. Guiducci, Private communication.

[64] A. Latina, Private communication.

[65] M. Morhac et al. Identification of peaks in multidimensional coincidence gamma-ray spectra. *Nucl. Instrum. Meth.*, A443:108–125, 2000.

[66] R. Makarov. Meeting on Beam Energy Measurements, DESY Zeuthen, June 06-08, 2007.

[67] T. Suzuki. General formulas of luminosity for various types of colliding beam machines. KEK-76-3, 1976.

[68] I. Will, P. Nickles and W. Sander. A laser system for the TESLA photo-injector, Internal Design Study. Max-Born-Institut, Berlin, 1994.

[69] User's manual of CAIN, Version 2.35.

[70] L. S. Brown and T. W. B. Kibble. Interaction of Intense Laser Beams with Electrons. *Phys. Rev.*, 133:A705–A719, 1964.

[71] B. Badelek et al. TESLA Technical Design Report, Part VI, Chapter 1: Photon collider at TESLA. *Int. J. Mod. Phys.*, A19:5097–5186, 2004.

[72] D. Yu. Ivanov, G. L. Kotkin and V. G. Serbo. Complete description of non-linear Compton and Breit- Wheeler processes. *Acta Phys. Polon.*, B37:1073–1077, 2006.

[73] M. V. Galynskii et al. Nonlinear effects in Compton scattering at photon colliders. *Nucl. Instrum. Meth.*, A472:267–279, 2001.

[74] A. Denner and S. Dittmaier. Complete $\mathcal{O}(\alpha)$ QED corrections to polarized Compton scattering. *Nucl. Phys.*, B540:58–86, 1999.

[75] M. L. Swartz. A complete order-α^3 calculation of the cross section for polarized Compton scattering. *Phys. Rev.*, D58:014010, 1998.

[76] R. Arnaldi et al. Performances of zero degree calorimeters for the ALICE experiment. *Nucl. Instrum. Meth.*, A456:248–258, 2001.

[77] ATLAS Collaboration. CERN/LHCC/2004-010 LHCC I-014, March, 2004.

[78] A. Seryi, L. Hendrickson and G. White. Issues of stability and ground motion in ILC, Contributed to 36th ICFA Advanced Beam Dynamics Workshop (NANOBEAM 2005), Kyoto, Japan, Oct 17-21 2005.

[79] J. Gervaise. High precision geodesy applied to CERN accelerators. *CERN 87-01*, pages 128–165, 1987.

[80] N. Muchnoi, Private communication.

[81] L. Keller, Private communication.

[82] H. Burkhardt et al. Halo and Tail Generation Computer Model and Studies for Linear Colliders. EUROTEV-REPORT-2008-076, 2008.

[83] S. P. Malton et al. Simulation of Beam Halo in CLIC Collimation Systems. EUROTEV-REPORT-2008-001, 2008.

[84] L. Deacon et al. Simulation Study of Laser-wires as a Post-linac Diagnostic for CLIC and ILC. EUROTEV-REPORT-2008-002, 2008.

[85] V. Gharibyan, N. Meyners and P. Schuler. Upstream Polarimetry with 4-Magnet Chicane, In the Proceedings of 2005 International Linear Collider Workshop (LCWS 2005), Stanford, California, Mar 18-22 2005.

[86] N. Muchnoi and M. Viti, Proposal in preparation.

[87] B. Aurand et al. Executive Summary of the Workshop on Polarization and Beam Energy Measurements at the ILC. arXiv:0808.1638, 2008.

Acknowledgments

The number of people I should thank for is large and it is difficult to record all of them in few lines. I think the first person I should thank is Jürgen Schreiber who offered me the possibility to have this wonderful experience here at DESY. He guided me since the beginning, indicating the path, supporting and motivating me.

I also want to thank Nickolai Muchnoi, who is not only a colleague but also a friend. Without his help, explanations and ideas I could not learn and understand so much as I did.

I thank Hermann Kolanoski for his precious suggestions and comments.

I want to thank Alexey Lyapin for the help and comments on the 3th and 5th chapters and Jörn Lange for the productive collaboration as summer student.

I want to thank DESY and the LC group for the friendly atmosphere who made working here a pleasant experience.

I am grateful to Martin for his patience, Ringo for the funny jokes, Andreas and Sergey for their help, Sabine, Wolfgang and Klaus to be always there, André for watching football in the evening, Karim, Sasha, Andry, Elena and everybody to be good friends.

A special thank is given to Ralph for his friendship, the sailing, the evenings in Berlin, for his help to prepare many figures that are involved in this thesis.

I would like to thank all the friends that I met here in Berlin, especially Marcello with him I spent the funniest moments.

A special thank is also directed to Moreno and Manuel who, although far away, remained wonderful friends.

Last but not least I thank my family who supported me every moment and Kerstin who was always close to me with her love.

Die VDM Verlagsservicegesellschaft sucht für wissenschaftliche Verlage abgeschlossene und herausragende

Dissertationen, Habilitationen, Diplomarbeiten, Master Theses, Magisterarbeiten usw.

für die kostenlose Publikation als Fachbuch.

Sie verfügen über eine Arbeit, die hohen inhaltlichen und formalen Ansprüchen genügt, und haben Interesse an einer honorarvergüteten Publikation?

Dann senden Sie bitte erste Informationen über sich und Ihre Arbeit per Email an *info@vdm-vsg.de*.

Sie erhalten kurzfristig unser Feedback!

VDM Verlagsservicegesellschaft mbH
Dudweiler Landstr. 99　　　　　　　Telefon　+49 681 3720 174
D - 66123 Saarbrücken　　　　　　　Fax　　　+49 681 3720 1749
www.vdm-vsg.de

Die VDM Verlagsservicegesellschaft mbH vertritt

Printed by Books on Demand GmbH, Norderstedt / Germany